JACK HI

Jack Higgins lived in ... e. Leaving school at fi... with the Royal Hors... East German borde... is subsequent emplo... as diverse as circus rousta... rk and, after taking an honours ... ogy and social psychology, teacher and ... y lecturer.

The Eagle Has Landed tu...ed him into an international bestselling author, and his novels have since sold over 250 million copies and have been translated into fifty-five languages. Many of them have also been made into successful films. His recent bestselling novels include *Bad Company*, *A Fine Night for Dying*, *Dark Justice*, *Toll for the Brave*, *Without Mercy* and *Storm Warning*.

In 1995 Jack Higgins was awarded an honorary doctorate by Leeds Metropolitan University. He is a fellow of the Royal Society of Arts and an expert scuba diver and marksman. He lives on Jersey.

Visit www.AuthorTracker.co.uk for exclusive updates on Jack Higgins.

Also by Jack Higgins

The Valhalla Exchange
To Catch a King
Dillinger
The Run to Morning
The Eagle Has Landed
A Prayer for the Dying
The Last Place God Made
Storm Warning
Day of Judgment
Solo
Luciano's Luck
Touch the Devil
Exocet
Confessional
Night of the Fox
A Season in Hell
Memoirs of a Dance-Hall Romeo
Cold Harbour
The Eagle Has Flown
Eye of the Storm
Thunder Point
On Dangerous Ground
Sheba
Angel of Death
Drink with the Devil
Year of the Tiger
The President's Daughter
Flight of Eagles
The White House Connection
Pay the Devil
Day of Reckoning
Edge of Danger
The Keys of Hell
Midnight Runner
A Fine Night for Dying
Bad Company
Dark Justice
Toll for the Brave
Without Mercy
East of Desolation

JACK HIGGINS

THE IRON TIGER

HARPER

This novel is entirely a work of fiction.
The names, characters and incidents portrayed in it are the work of the author's imagination. Any resemblance to actual persons, living or dead, events or localities is entirely coincidental.

Harper
An imprint of HarperCollins*Publishers*
77–85 Fulham Palace Road,
Hammersmith, London W6 8JB

www.harpercollins.co.uk

This paperback edition 2008
1

First published in Great Britain by
John Long 1966

A catalogue record for this book is available from the British Library

ISBN 978-0-00-785193-5

Set in Sabon by Palimpsest Book Production Limited,
Grangemouth, Stirlingshire

Printed and bound in Great Britain by
Clays Ltd, St Ives plc

PUBLISHER'S NOTE

THE IRON TIGER was first published in the UK by John Long in 1966, then later by Signet books, but has been out of print for some years.

In 2008, it seemed to the author and his publishers that it was a pity to leave such a good story languishing on his shelves. So we are delighted to be able to bring back THE IRON TIGER for the pleasure of the vast majority of us who never had a chance to read the earlier editions.

For Brenda Godfrey who likes a good story

Foreword

India has always fascinated me, although I had not visited the country when I wrote *The Iron Tiger*. I have since, of course, and was delighted to find that, thanks to careful research, I had got it right. The period during which I wrote the book, the early sixties, was one in which the Chinese occupying power treated the Tibetan people with great brutality and many thousands of those unfortunates died. The Chinese invasion of India hardly made them popular in that country and, because of this, *The Iron Tiger* was a great success with Indian people. However, a strange thing happened. It made me, as the author, highly popular for a while as for some reason people believed that I had simply fictionalised a true story about myself and that the events of the book had actually taken place.

Jack Higgins
July 1996

1

The Place of Silence

Beyond the mountains, the sky was sapphire and blue, a golden glow spreading across the ice caps as the sun slowly lifted. Below, the valleys lay dark and quiet, the only sound the tiny, insignificant drone of the Beaver's engine as it followed the maze through to Tibet.

Jack Drummond was tired and a slight dull ache behind his right eye nagged constantly. Too many late nights, too much whisky and he was getting old. Too old to be dicing in the worst flying area in the world at sixteen thousand feet in a non-pressurised cabin.

He turned to Cheung and grinned. 'There's coffee in a black flask under your seat. I could do with some.'

His companion was Chinese, but it was obvious that he had European blood. The eyes were startlingly blue in the bronzed, healthy face and his mouth lifted slightly in a quirk of ironic good humour.

He wore a heavy sheepskin coat and an astrakhan cap and shivered as he opened the vacuum flask and poured coffee into a plastic cup.

'Is it always as cold as this?'

Drummond nodded. 'The wind comes all the way from Mongolia. There have been times when it's stripped pieces off the fuselage.'

Cheung peered down into the jagged valley below. 'What would happen if the engine stopped?'

Drummond laughed harshly. 'You're joking, of course.'

Cheung sighed. 'It becomes clearer minute-by-minute that you have been earning your money during the past six months.'

'And perhaps a little more?'

The Chinese smiled amiably. 'My dear Jack, in Formosa, we subsist almost entirely on the goodwill of our American friends. If it wasn't for their generosity, we couldn't even afford such minor gestures as this Tibetan venture.'

Drummond shrugged. 'It doesn't worry me. A couple more trips and I'm through. I've done this run too often. I'm on borrowed time.'

Cheung frowned. 'But Jack, there is no one else. What will we do?'

'There's always someone else,' Drummond said. 'You'll find him in one bar or another in Calcutta. Plenty of ex-R.A.F. types who can't settle down or the other kind who've lost their licences to fly commercially. They'll go anywhere if the money's right.'

They moved on through a landscape so barren that it might have been the moon, great snow-covered peaks towering on either side. Drummond handling the plane with the skill of genius. Once they dropped sickeningly in an air pocket, and on another occasion flew along a canyon so narrow that the wingtips seemed to brush the rock face. Finally, they lifted across a snow-covered ridge and plunged into space.

Beneath them an enormous valley dropped ten thousand feet, black with depth, purple and gold, great shimmering banks of cloud strung across it in broken strands. Perhaps seven or eight miles away on the other side was the last frozen barrier between Balpur and Tibet.

The sound of the engine suddenly seemed strangely muted and Cheung sighed through the uncanny quiet. 'The most beautiful sight I've ever seen.'

'The Place of Silence, that's what they call it,' Drummond told him. 'Used to take two days to get across on foot when caravans were still coming through.'

The Beaver seemed to glide on through the enormous blue vault, drifting through the shadows, and then they burst out into golden sunlight and the final barrier rose before them.

Drummond eased back the stick and the Beaver lifted, the sound of the engine deepening into a full-throated roar and a deep valley appeared between the peaks.

'Sangong Pass,' he called above the roar of the engine.

They swept into the pass, a brilliant red and gold leaf, bright against the dark walls, and the frozen earth rose to meet them. Drummond gave the Beaver full power and pulled the stick right into the pit of his stomach.

Cheung held his breath, waiting for the crash as they rushed to meet the skyline, wheels no more than ten feet above the boulder-strewn ground and then they were over the hump and flashing across a great, cold glacier.

Rolling steppes, golden in the morning sun, stretched to the horizon and Drummond grinned. 'Now you know why I charge two thousand a trip.'

Cheung wiped sweat from his forehead with the back of a gloved hand and managed a weak smile. 'I'm beginning to get the point. How much further?'

'Ten or twelve miles, that's all. Better get ready.'

The Chinese reached behind his scat for a sub-machine gun, cocked it and held it across his knees. As the Beaver descended, he could see a narrow river, brawling across a mass of tumbled boulders, widening into a shallow lake. A hundred yards to the left, sheltered against a rock escarpment, was a ruined monastery, a scattering of houses at its feet.

Drummond pointed to a wide sand flat at the far end of the lake. 'That's where we land if we get the signal.'

'And if not?'

'We get the hell out of here.'

He circled, coming in low across the lake, and Cheung pointed excitedly. 'There are people down there, standing in the shallows.'

'Women doing the washing,' Drummond said and swung in across the village, turning away from the escarpment and the fire-blackened ruins of the monastery.

'What happened there?' Cheung demanded.

'It was a headquarters for local resistance back in 1950 when the Chinese Reds first invaded Tibet. There was a siege for a couple of days, but it didn't last long. They brought up a couple of field guns and blew the necessary holes through the walls.'

'Then what?'

Drummond shrugged. 'They saved everything worth having, then burnt the place to the ground and executed a couple of hundred monks.'

'To encourage the others?'

Drummond nodded and took the Beaver round to the other side of the lake in a graceful curve. 'Not that it's done them much good. In areas like this, they only control the towns.'

He took the Beaver down towards the village again and Cheung touched his arm quickly. 'Is that the signal?'

Three flares, spaced out in a crude triangle, started to burn furiously, plumes of white smoke lifting into the cold air and Drummond nodded.

He throttled back, turned the Beaver into the wind, dropped it neatly down on the firm, sandy shore of the lake and taxied along to the far end.

The women washing clothing in the shallows a few yards away, moved up on the shore, their long woollen *shubas* tucked into their belts and stood in a tight little group, watching the plane.

Cheung reached for the door handle and Drummond shook his head. 'Not yet. We've got to be sure.'

At that moment, a horseman galloped over the crest of the slope above them and plunged down towards the plane. Drummond switched off the engine and grinned in the sudden silence.

'There's our man.'

As he opened the door and jumped to the ground, the rider reined in his small Tibetan horse, dismounted and strode towards them. He was tall and muscular with a deeply-tanned face and high Mongolian cheekbones. He wore a long, wide-sleeved robe and sheep-skin *shuba* which left his chest bare, and knee-length boots of untanned hide. His hair was coiled into plaits and he wore a sheep-skin hat.

'His English isn't much good,' Drummond said to Cheung as the Tibetan approached. 'We'll use Chinese and for God's sake treat him with respect. He's a nobleman. They can be touchy about things like that.'

The Tibetan grinned and held out his hand,

and behind him another dozen men rode down to the shore. 'It is good to see you again, my friend. You have more guns for us?'

Drummond nodded as he shook hands. 'Your men can unload them as soon as they like. I don't want to hang around here for any longer than I have to.'

The Tibetan shouted an order and he and Drummond and Cheung moved out of the way. 'Moro, this is Mr Cheung,' Drummond said. 'He's the Balpur representative of the Chinese National Government, the people who've been supplying the guns and ammunition I've been flying in to you during the past six months.'

Moro took Cheung's hand warmly. 'Before the Lord Buddha brought the way of peace to this land, the Tibetans were warriors. Your guns have helped us prove to the Communists that we can be warriors again. You will take tea with me before you leave?'

Cheung turned to Drummond. 'Have we time?'

'I don't see why not.' Drummond offered

the Tibetan a cigarette. 'Any Reds in the area recently?'

'One patrol,' Moro said. 'Fifteen men. They turned up a week ago.'

'What happened?'

The Tibetan grinned. 'You'll see when we reach the village.'

They went over the escarpment and walked towards the houses, the Tibetan with the bridle of his horse looped over one arm.

'Mr Cheung has to make a special report to his government in Formosa about the state of things here,' Drummond said. 'He thought he'd like to see for himself.'

'How strong are the Reds in this area?' Cheung asked.

'Their nearest real strength is at a town called Juhma about a hundred miles from here,' Moro said. 'Half a regiment of infantry. No more than four hundred men. At larger villages like Hurok which is thirty miles east across the plain, they keep a cavalry troop. Between the villages they are as nothing.'

'There have been no large scale troop

movements, no road building in this area of the border at all?'

'Not here, but further east towards the Aksai Chin and the Ladakh where they fought the Indians in 1962, they have built many roads.' The Tibetan looked surprised. 'Why would they need roads here?'

'They have claimed Balpur,' Cheung said simply.

Moro laughed, showing strong white teeth. 'They have claimed the whole world, is this not so?'

'They came to the outskirts of the village, a small, mean place, single-storey houses of mud and wattle strung along either side of the single street.

Several children ran forward excitedly and followed them, keeping a respectful distance from Moro who occasionally flicked out with the plaited leather whip that hung from his left wrist as someone moved too close.

They came to a house near the centre of the village that seemed larger than the others and he opened the heavy wooden door and led the way in.

There were no windows and in the half-darkness Drummond was aware of the mud walls, the sheepskins on the floor. On a stone hearth in the centre, a fire of yak dung burned brightly and an old Tibetan woman was crumbling brick tea into a cauldron of boiling water. She added butter and a pinch of salt and the men squatted on a sheepskin beside the fire.

They waited in silence for the tea as ritual demanded. The old woman filled three metal cups and gave them one each. Moro took a sip, nodded in approval and they drank.

It was, as always, curiously refreshing and Drummond held out his cup for more. 'How are things going generally?'

Moro shrugged. 'They will not be beaten here, we cannot hope to accomplish so much, but we can keep them occupied, make life difficult.'

'What about arms?' Cheung said. 'You need more?'

'Always more. We can't fight them with broadsword and musket.'

'You were going to tell us about the patrol,' Drummond reminded him.

Moro nodded and got to his feet. 'I was forgetting. If you have finished your tea, I will show you now.'

They moved into the street, blinking in the bright, clear morning sunlight and the Tibetan led the way through the crumbling houses, the small tail of children keeping pace with them.

The great wooden gates in the outer wall of the monastery swung crazily from their hinges, half-burnt away and blackened by fire.

They crossed the courtyard beyond, still followed by the children, and mounted the broad steps to the ruin of what had once been one of the most famous seats of learning in Western Tibet.

The doors had disappeared, splintered into matchwood by high explosive shells, and inside bright sunlight streamed down through holes in the roof.

'There was a library here,' Drummond told Cheung. 'It held more than fifteen thousand books and manuscripts, most of them over a thousand years old. The Chinese burned the lot quite deliberately.'

Beyond, in the shadows, something stirred and a kite rose lazily into the air, great ragged black wings brushing the roof beams and Drummond was aware of Cheung's breath hissing between his teeth sharply.

Disturbed by the bird's passage, something was swinging to and fro, half-in, half-out of the bright shafts of sunlight cutting down through the darkness.

Drummond moved a little closer. It was a Chinese soldier, swinging by a rope from one of the charred beams, tongue protruding obscenely from the black, swollen face. Where the eyes had been, were only empty, ragged sockets and one ear had been torn off.

As his eyes became accustomed to the half-light, he saw the others, each hanging from a beam, staring blindly into eternity.

'We were away when they arrived,' Moro said simply. 'When we returned, the fools were so busy ravishing the women, they had not even thought to post a guard.'

One of the children ran forward with a harsh laugh and grabbed the nearest corpse by the legs, swinging it from side to side furiously and the other children followed suit, running through the shadows, dodging the swinging bodies, helpless with laughter.

Drummond turned and moved into the sunlight again, his mouth dry. 'I think we should be making a move.'

Mr Cheung didn't speak. His face was strangely pale and there was shock and pain in his eyes as they returned to the village. Moro whistled for his horse, caught the bridle and led the way back to the lake.

'What did you bring this time?' he asked Drummond.

'Automatic rifles, sub-machine guns and ten thousand rounds of ammunition.'

The Tibetan nodded. 'Good, but we could do with some explosive next time.'

Drummond glanced at Cheung enquiringly. 'Can you manage that?'

The Chinese nodded. 'I think so. Would a fortnight today be too soon?'

'Not for me,' Drummond said. 'Two more trips and I'm finished. The sooner I get them done, the better I'll like it.'

'A fortnight, then,' Moro said and they

went over the escarpment and down to the shore beside the lake.

His men had unloaded the plane and already several packhorses were on their way to the village. Drummond gave him a final cigarette, climbed in and strapped himself into his seat. As the engine roared into life, Mr Cheung turned and held out his hand.

'We are united in the same struggle,' he said and climbed into the plane.

As he closed the door and fastened his seat belt, the Beaver turned into the wind and started to taxi along the shore, sand whipped up by the propeller rattled against the windows. A moment later, the bluff at the far end of the lake was rushing to meet them and they were rising into the air.

Drummond circled once and Moro, already back in the saddle, waved, turned his horse and galloped back towards the village.

Drummond checked his instruments and started to gain altitude. 'Well, what did you think?'

'Words fail me.'

'I thought they would.'

Cheung lit a cigarette and sighed heavily. 'To you, it is nothing. Jack. Dangerous, unpleasant, yes, but something you are mixed up in for one reason only – money.'

'And to you it's a holy war,' Drummond said. 'I know, only don't start trying to get me to join the crusade. I had a bellyful of that kind of thing in Korea. Enough to last a lifetime.'

'All right,' Cheung said wearily. 'What about these explosives Moro wants on the next trip? If I have them delivered to the rail-head at Juma by next weekend can you pick them up?'

'I'm flying down tomorrow with Major Hamid,' Drummond said. 'He's taking a week's leave. He thought he might enjoy it more if I went along. Why don't you join us?'

Cheung shook his head. 'I'd like to, but I've been getting behind with the paperwork and I'm supposed to be dining with the old Khan on Saturday night.'

'Suit yourself,' Drummond said.

Another two thousand. That brought the

total standing to his credit in the Bank of Geneva to £23,000. Two more trips plus the money Ferguson owed him and he'd have a straight £30,000. After that, he was finished. Time he had a rest. He leaned back in the seat, humming to himself and concentrated on his instruments as he took the Beaver slanting across the glacier and into the pass.

Moro galloped alongside the packhorses, whistling, slashing their bony rumps with the heavy leather riding whip. He urged his mount forward and entered the village first, clattering over the loose stones and dismounted outside his house.

The children had disappeared and the street was quite deserted as he stood there listening to the sound of the Beaver in the distance, drawing happily on the English cigarette Drummond had given him.

Doors opened in the houses along the street and one by one soldiers emerged in peaked caps and drab quilted jackets. As Moro turned, the door to his house opened and a young officer emerged. He wore a beautifully tailored riding coat with fur collar and the red star of the Army of the People's Republic gleamed brightly in his cap.

'I did well?' Moro said.

The young officer took the English cigarette from the Tibetan's lips and inhaled deeply. A sunny smile appeared on his face.

'Excellent. Really quite excellent.'

Moro nodded, the eager smile still firmly in place and together they stood there, listening to the sound of the Beaver fade into the pass.

2

House of Pleasure

Drummond emerged from the hot room, dropped his towel on the tiled floor and dived into the plunge bath, swimming down to touch the brightly coloured mosaic face of Kali, the Great Mother, staring blindly into eternity through the green water as she had done for a thousand years.

He surfaced and one of the house girls moved out of the steam and squatted at the side of the ancient bath, holding a tray containing a slender coffee pot and tiny cups. Drummond swam towards her and she handed him down a cup as he floated there in the water.

She was like all the rest of them, startlingly

beautiful, with delicate features and great kohl-rimmed eyes. Her green silken sari was saturated with steam, outlining to perfection the firm body, the curving breasts.

As Drummond sipped his coffee, he heard a harsh laugh somewhere near at hand and Hamid's great voice boomed between the walls. He was singing the first stanza of *Zukhmee-Dil*, a ballad immensely popular on the North-West Frontier, at one time the favourite march of the Khyber Rifles.

Wullud sureen shuftauloo-maunind duryah,
Ufsose! mun n'shinnah.

Drummond handed his cup back to the girl and threw the song back at the Pathan, translating into English.

There's a boy across the river with a
bottom like a peach, but, alas, I cannot
swim.

Hamid bellowed with laughter as he moved out of the steam, a towel about his waist. He was a Pathan of the Hazara tribe, dark-skinned, bearded. A handsome buccaneer of a man of six feet three with broad muscular shoulders.

He smiled hugely. 'Feeling better, Jack, headache gone?'

'Ready for anything,' Drummond replied.

'Me, too.' The Pathan ran his fingers through the long hair of the girl who still squatted at the side of the bath. 'A good song, that, but where love's concerned, I'm the old-fashioned kind.'

He pulled the girl to her feet and the damp sari parted exposing her left breast. 'Now there's a thing.' He swung her up into his arms and grinned down at Drummond. 'See you later.'

Drummond swam lazily across to the other side of the bath and back again. He repeated the process twice and then hauled himself out over the stone edge, smoothed by time. He picked up his towel, wrapped it around his waist and padded across the warm tiles.

The next room was long and narrow with a vaulted roof and lined with cubicles, some with curtains drawn. From one he heard Hamid's deep chuckle followed by the lighter laughter of the girl and smiled to himself.

He went into the end cubicle, pulled the bell cord in the corner, climbed on to the stone massage slab and waited. After a while, the curtain was drawn and Ram Singh, the proprietor, entered followed by several bearers carrying buckets of hot and cold water.

The Hindu smiled. 'All is in order, Mr Drummond?'

'You've made a new man of me,' Drummond said. 'We could do with you in Sadar.'

The Hindu rolled his eyes to heaven in simulated horror. 'The end of the world, Mr Drummond. The end of the world. I will send Raika.'

He withdrew and Drummond lay there staring up at the ceiling. *The end of the world*. Well, that wasn't far off as a description of Sadar. A capital city with a population of three thousand, which gave some idea of the size of Balpur itself. A barren, ugly land, harsh and cruel as its inhabitants. The last place God made. Well, not for much longer, Praise be to Allah.

The curtain rustled and when he turned his head, Raika had entered. She was strikingly beautiful and wore a ruby in one nostril and great silver ear-rings with little bells on the end that tinkled when she moved her head.

Her sari was of blue silk threaded with gold and outlined every curve of her graceful body. Drummond nodded, and without speaking she started to work.

First came the hot rinse, water so scalding that he had to stifle the cry of pain that rose in his throat. She worked on his limbs to start with, first with the brushes and then with practised hands, loosening taut muscles, relaxing him so completely that he seemed to be floating, suspended in mid-air.

And as always, he was amazed at the matter-of-factness of it all, the lack of overt sensuality. But then this was India where life and death, love and the flesh, were all a part of one great mystery.

She sluiced him down again with another

bucket of hot water that was followed immediately by one so cold it drew the breath from his body. He gasped and there was a glint of laughter in her eyes, barely contained, so that at once she became real, a creature of flesh and blood.

She leaned over him, the damp sari gaping to the waist and Drummond cupped a hand over one sharply pointed breast. She went very still and stayed there in that position, leaning across him, her hand still reaching for the brush.

Drummond stared up at her, the nipple hardening against his palm and something stirred in her eyes. Her head came down slowly, the mouth slightly parted, and as he slid his free hand up around her neck, there was a discreet cough at the entrance.

Raika stood back at once completely unconcerned, and Drummond sat up. Ram Singh peered through the curtain, an anxious frown on his face.

'So sorry, Mr Drummond, but there is a person to see you.'

Drummond frowned. 'A person?'

'A Miss Janet Tate.' Ram Singh laughed nervously. 'An American lady.'

'In this place?'

Hamid appeared at the Hindu's shoulder, a cigarette in his mouth. 'A day for surprises, Jack. Any idea who she is?'

'There's one way of finding out.'

Drummond tightened the towel around his waist, left the cubicle and went into the next room. It was beautifully furnished with heavy carpets, low divans and round brass coffee tables at which several clients were relaxing after the rigours of the bath.

He crossed the room followed by Hamid and the Hindu, knelt on a divan and peered through the latticed partition of wrought iron into Ram Singh's office.

Janet Tate stood at the desk, examining a figurine of a dancer. She put it down, turned and looked around her with interest, moving very slowly across the floor, incredibly lovely in the yellow dress, the long, shoulder-length black hair framing her calm face.

Hamid sighed softly. 'A houri from Paradise itself, sent to delight us.'

Drummond straightened, a frown on his face. 'Get me a robe, will you?'

The Hindu was back in a moment and helped him into it. 'Aren't you going to dress first?' Hamid said.

Drummond grinned. 'My curiosity won't allow me to wait that long.'

When he opened the door and stepped into the office, Janet Tate was examining a tapestry hanging on one wall. She turned quickly and stood quite still.

The man who faced her was about forty, the crisp black hair already greying a little at the temples. He was perhaps six feet in height, well built with good, capable hands. She noticed them particularly as he fastened the belt of his robe.

But it was the face that interested her, the slight ironic quirk to the mouth of someone who laughed at himself and other people too much; the strong, well-defined bones of the Gael. Not handsome, the ugly, puckered scar running from the right eye to the corner of the mouth had taken care of that, but the eyes were like smoke slanting across a hillside on a winter's day and she was aware of a strange, inexplicable hollowness inside her.

'Mr Drummond? I'm Janet Tate.'

She didn't hold out her hand. It was as if she was afraid to touch him, afraid of some elemental contact which, at this first moment, she might be unable to control.

And then he smiled, a smile of such devastating charm that the heart turned over inside her. He shook his head slowly. 'You shouldn't have come here, Miss Tate. It's no place for a woman.'

'That's what the man at the hotel told me,' she said. 'But they have girls here. I saw two as I came in.'

And then she realised and her eyes widened. Drummond helped himself to a cigarette from a sandalwood box on the desk. 'What can I do for you?'

'I'm trying to get to Sadar. I believe you might be able to help.'

He frowned his surprise. 'Why on earth do you want to go to Sadar?'

'I'm a nurse,' she said. 'I've been sent here by the Society of Friends to escort the Khan of Balpur's young son to our Chicago hospital. He's to undergo serious eye surgery there.'

And then Drummond remembered. Father Kerrigan had told him about it before leaving. But the old priest had said they were expecting a doctor.

'So you're a Quaker.'

'That's right,' she said calmly.

'First visit to India.'

She nodded. 'I've just finished a two-year tour of service in Vietnam. I was on my way home on leave anyway, so the Society asked me to make a detour.'

'Some detour.'

'You can take me?'

Drummond nodded. 'No difficulty there. I fly a Beaver, there's plenty of room. Just one other passenger – Major Hamid, Indian Army adviser in Balpur, not that they have much of an army for him to advise. We'll take off about four-thirty, make an over-night stop at Juma and fly on through the mountains to Sadar in the morning. Much safer that way.' He crushed his cigarette into the

Benares ashtray. 'If you'll hang on, I'll go and dress.'

He started for the door and she said quickly, 'I was forgetting, I have a message for you from a Mr Ferguson.'

When he turned, it was the face of a different man, cold, hard, wiped clean of all expression, the eyes like slate.

'Ferguson? Where did you meet Ferguson?'

'On the train from Calcutta. He was very kind to me. He wants you to call on him at the usual place before you leave.' She smiled brightly. 'It all sounds very mysterious.'

An invisible hand seemed to pass across his face and he smiled again. 'A great one for a joke, old Ferguson. I shan't be long.'

He left her there and hurried through the other room to the changing cubicles where he dressed quickly in a cream nylon shirt, knitted tie and single-breasted blue suit of tropical worsted.

When he returned to the office, Hamid was sitting on the edge of the desk, Janet in the chair beside him looking up, a smile on her face.

Drummond was aware of a strange, irrational jealousy as he moved forward. 'I see Ali's managed to make his own introductions, as usual.'

'If I must be formally introduced, then I must.' Hamid grinned down at Janet. 'Jack was at one time a Commander in the Navy. He's never got over it. They're very correct, you know.'

He jumped to his feet and stood there waiting for Drummond to speak, a handsome, challenging figure in his military turban and expertly tailored khaki drill uniform, the medal ribbons a bright splash of colour above his left breast pocket.

Drummond sighed. 'Trapped, as usual. Miss Janet Tate, may I present Major Ali Mohammed Hamid, D.S.O., a British decoration, you'll notice. Winchester, one of our better public schools, and Sandhurst. Rather more class than West Point, don't you think?'

Hamid took her left hand and raised it to his lips gallantly. 'See how the British have left their brand on us, clear to the bone, Miss Tate?'

'Don't look at me,' Drummond said. 'I'm a Scot.'

'The same thing,' Hamid said airily. 'Everyone knows it's the Scots who rule Britain.'

He gave his arm to Janet and they moved out into the bright, hot sunshine. Across the square, there was a low wall and beyond it, the river, usually two miles wide at this point, but as always when winter approached, narrowing to half a mile or less, winding its way through endless sandbanks.

'Is this still the Ganges?' Janet asked.

'Ganges, Light amid the Darkness, Friend of the Helpless. It has a thousand names,' Hamid said as they strolled towards the low wall. 'To bathe in her waters is to be purified of all sin, or so the Hindus believe.'

Janet leaned on the wall and looked down the cobbled bank into the in-shore channel at the brown, silt-laden water. 'It looks pretty unhealthy to me.'

Drummond lit a cigarette and leaned beside her. 'Strangely enough, it does seem to have health-giving properties. During religious festivals pilgrims drink it, often at places where the drains disgorge the filth of the town, but they never seem to suffer. Bottled, it keeps for a year. They say that in the old days when taken on board clipper ships in Calcutta, it outlasted all other waters.'

Down below at the river edge some kind of ceremony was taking place and she glanced up at Hamid. 'Can we go down?'

'But of course. Anything you wish.'

'Not me,' Drummond said. 'If I'm going to see Ferguson before we leave, I'd better be moving.' He glanced at his watch. 'It's almost two o'clock now. I'll see you back at the hotel at four.'

He moved away across the square quickly and Janet watched him go, a slight frown on her face. 'I believe Mr Ferguson said he was in the tea business.'

'That's right,' Hamid said. 'Jack has an air freight contract with him. Ferguson usually comes up to see him once a month. He has a houseboat lower down the river from here.'

'You said Mr Drummond was once a naval commander?'

'Fleet Air Arm.'

'He was a regular officer, then? He would have been too young to have been a full commander during the war.'

'Quite right.' The Pathan still smiled, but there was a slight, cutting edge to his voice, a look in the eye that warned her to go no further. 'Shall we go down?'

They stood on the edge of a small crowd and watched the ceremony that was taking place. Several people stood knee-deep in the water, the men amongst them stripped to the waist and daubed with mud. One of them poured ashes from a muslin bag into a larger paper boat. Another put a match to it and pushed the frail craft away from the bank, out into the channel where the current caught it. Suddenly, the whole boat burst into flames, and a moment later sank beneath the surface.

'What were they doing?' Janet asked.

'The ashes were those of a baby,' Hamid said. 'A man-child because the ceremony is expensive and not worth going through for a girl.'

'And they do this all the time?'

He nodded. 'It is every Hindu's greatest dream to have his ashes scattered on the waters of Ganges. Near here there is a *shamsan*, a burning place for the dead. Would you like to see it?'

'Do you think I can stand it?'

He smiled down at her. 'Two years in Vietnam, you said. If you can take that, you can take anything.'

'I'm not so sure.' She shook her head. 'India's different, like no other place on earth. Ferguson told me that and he was right.'

As they moved along the shore, she could smell woodsmoke, and up ahead there was a bullock cart, three or four people standing beside it.

As they approached, she gave a sudden gasp and moved closer to Hamid. A naked man was lying on a bed of thorns, eyes closed, his tongue protruding, an iron spike pushed through it. His hair and beard were matted and filthy, his body daubed with cowdung and ashes.

'A *saddhu*, a holy man,' Hamid said, throwing a coin into an earthenware jar that stood at the man's head. 'He begs from the mourners and prays for the souls of the dead.'

There was nothing to distinguish the place from any other stretch of the shore, no temples, no monuments. Only the ashes of old fires, the piles of calcined bones and here and there a skull, glaring blindly up at the sky.

The people by the fire laughed and joked with each other and as the flames roared through the criss-crossed logs of the funeral pyre in a sudden gust of wind, she caught the sweetly-sick, distinctive stench of burning flesh and her throat went dry, panic threatening to choke her.

She turned, stumbling against Hamid, and beyond him in the water something turned over in the shallows, a rotting body, arms trailing, a grey headed gull swooping down, beak poised to strike.

There was immediate concern on his face, and unconsciously he used her first name. 'Janet, what is it?'

'The smell,' she said. 'Burning flesh. I was in a village called Nonking north of Saigon last year. The Viet Cong made one of their night raids and set fire to the hospital.' She stared back into the past, horror on her face. 'The patients, we could only get half of them out. There are nights when I can still hear the screams.'

She was aware of his hand under her arm and they were climbing rapidly up the bank, across a narrow stone causeway. Suddenly, they moved into a different world, a place of colour and light, scarlet hibiscus and graceful palms.

They walked through trees along a narrow path and emerged on to a stone, loopholed terrace high above the river, a couple of ancient iron cannon still at their stations as they had been for three hundred years.

Hamid pushed her gently forward. 'And behold, said the genie . . .'

She gave an excited gasp and leaned across the wall. Between the sandbanks, hundreds of flamingoes paced through the shallows, setting the very air alight with the glory of their plumage. Hamid picked up a stone and tossed it down, and immediately the sky was filled with the heavy, pulsating beat of their wings as they lifted in a shimmering cloud.

He looked down at her gravely. 'Back there, death, Janet. Here, life in all its magnificence. They are both sides of the same coin. This you must learn.'

She nodded slowly and slipped her hand into his arm. Together, they walked back quietly through the trees without speaking.

Beyond the old quarter of the town, Drummond moved into an area of stately walled villas and beautiful gardens, the homes of rich merchants and government officials. A narrow path, fringed with eucalyptus trees, brought him to the river bank again.

A red houseboat was moored at the end of an old stone wharf about forty yards away and Ferguson's Sikh bearer squatted on the cabin roof. When he saw Drummond, he scrambled to the deck and disappeared below.

Drummond crossed the narrow gangway and stepped on to the deck which had been scrubbed to a dazzling whiteness. Several cane chairs and a table were grouped under an awning at the stern and as he sat down, the Sikh appeared with a tray containing a bottle of gin, ice-water and glasses. He placed the tray on the table and withdrew without speaking.

Drummond helped himself to a drink and walked to the stern rail, staring out across the river and thinking about Janet Tate, as a boat slipped by, sail bellying in the breeze.

There was a clink of a bottle against glass and when he turned, Ferguson was sitting at the table, pouring himself a drink.

'You're looking fit, Jack. Nothing like a steam bath to pull a man round after a hard night.'

'Hullo, Fergy, you old rogue,' Drummond said. 'I got your message. It was delivered in person at Ram Singh's House of Pleasure by a rather delectable little Quaker girl in a yellow dress.'

'God in heaven,' Ferguson said, astonishment on his face. 'She didn't, did she?'

'I'm afraid so.' Drummond sat down and took a cheroot from an old leather case. 'Her first visit to India, apparently. She's a lot to learn.'

'I found her travelling from Calcutta second class,' Ferguson said. 'Can you imagine that? What's all this about the Khan's son needing eye surgery?'

'The boy fell from his horse a month ago and took a nasty knock. The sight started to fail in the right eye, so the old man had me fly a specialist up from Calcutta. He's got a detached retina and his balance has been affected.'

'Tricky surgery to put that right.'

'It seems the big expert's on the staff of some Quaker foundation hospital in Chicago. Father Kerrigan got in touch with them and they agreed to take the case. Said they'd send a doctor to escort the boy.'

'Instead, you get Janet Tate.'

'Who was already in Vietnam and due home on leave, so they saved on the fare.' Drummond grinned. 'Never look a gift horse in the mouth, Fergy.'

Ferguson frowned slightly. 'She's a nice girl, Jack. A hell of a nice girl. I wouldn't like to see her get hurt.'

'So?' Drummond said coolly.

Ferguson sighed. 'All right, let it go. What have you got for me this time?'

Drummond took several spools of film from his pocket and pushed them across. 'That's the lot. You've got the whole Balpur-Tibet border region now.'

'You've finished?'

Drummond nodded. 'Trip before last. A good job, too. Cheung decided to fly in with me on the last trip, so I couldn't have set the camera up if I'd wanted to.'

Ferguson smiled and shook his head. 'Our Nationalist friends are still at it, are they? I wonder what Washington would say if they knew?'

'I couldn't care less,' Drummond said. 'A couple more trips and I'm through. I've told Cheung that already.'

Ferguson applied a match to the bowl of his old briar pipe and coughed as the smoke caught at the back of his throat. 'How did you find things last trip? Any signs of Chinese activity?'

'Swinging on the end of a rope,' Drummond said. 'Moro and his band dealt with a cavalry patrol in their own inimitable fashion, that's all.'

'Nothing else? You're sure about that?'

Drummond nodded. 'Moro says that all the activity's still in the Aksai Chin, Ladakh region. No sign of any large scale interest in the Balpur border area at all.'

'That's strange, you know. They've claimed it officially and the brutal truth is they're on pretty firm ground this time, historically speaking.'

'They can have it, for all I care,' Drummond said. 'Another month, and I'm out.'

Ferguson poked a match into the end of his pipe to clear the air hole and said casually, 'What were you thinking of doing?'

'Nothing you'd be interested in. I'm finished, Fergy. I've had enough. How long have I given you now; four years, five? I've played this sort of game on every border from Sarawak to Kashmir. I can't go on forever. Nobody can.'

'You've done a good job, Jack. I'm not denying that,' Ferguson said. 'But you've been well paid.'

'What about last year when the Indonesians shot me down in Borneo?' Drummond reminded him. 'They chased me through that jungle for three weeks before I managed to scramble across the border.' He ran a finger down the ugly scar that stretched from his right eye to the corner of his mouth. 'I spent a month in hospital and what happened? You paid me the same as always. No more, no less.'

Ferguson sighed, took an envelope from his pocket and pushed it across. 'Three thousand, deposited as usual with your Geneva bankers. You know how to get in touch with me if you change your mind.'

'That'll be the day.' Drummond opened the envelope, examined the deposit slip, then put it in his wallet. 'It's been fun, Fergy.'

He moved along the deck to the gangplank and stepped on to the wharf. 'One more thing, Jack,' Ferguson called. 'Don't forget who the Beaver belongs to when you've finished up there. Government property, you know.'

'And just how would you like to set about proving that?' Drummond said and started to laugh as he walked away along the wharf.

3

The Nightwalkers

Janet stepped out of the shower, dried herself quickly and went into the bedroom, the towel wrapped around her slim body. The window to the terrace was open and she stood in the shadows and looked out.

A bank of cloud rolled away from the moon and Juma was bathed in a hard white light, flat-roofed houses straggling down to the river below. The night sky was incredibly beautiful with stars strung away to the horizon where the mountains lifted uneasily to meet them.

It was peaceful and quiet, a dog barking hollowly somewhere in the night. In the streets below, she could see torches flaring and then a drum started to beat monotonously, joined a moment later by some stringed instrument, and the sound of laughter drifted up on the warm air.

There was a discreet tap on the door and she called quickly, 'Who is it?'

'Ali – can I speak to you for a moment?'

She pulled on her dressing gown, fastened the cord and opened the door. Hamid came in, resplendent in his best uniform.

'How are you feeling?'

'Fine. I slept for an hour, then had a shower.'

'Good.' He hesitated and then went on apologetically. 'I'm sorry about this, Janet, but I'm afraid I'd already arranged something for this evening.' He glanced at his watch. 'As it is, I'm pressed for time.'

'A lady?'

'I hope not,' he said solemnly.

She chuckled. 'You're quite incorrigible. Better not keep her waiting.'

'Jack went out to the airstrip to check on some cargo we're taking with us tomorrow. Motor spares, I think. He shouldn't be more than half an hour.'

She listened to the sound of his footsteps fade along the narrow passage and then closed the door. She stood with her back to it, a slight frown on her face and then walked slowly across to the window.

The drumming was louder now, an insistent throbbing that filled the night and someone was singing in a high, reedy voice, hardly moving from one note to another, monotonous and yet strangely exciting.

She hurried across to the bed, opened her second suitcase and took out a sleeveless black dress in heavy silk that she had purchased in a moment of weakness in Saigon. She held it against herself for a moment in the mirror, and then smiled and started to dress. When she was ready, she pulled on a white linen duster coat against the night air, wound a silk scarf around her head and went downstairs.

The Hindu night clerk dozed at his desk, but came awake at once when she touched him lightly on the shoulder. 'I want to go to the airstrip. Can you get me a *tonga*?'

'Certainly, memsahib. Come this way.'

He took her out through the entrance and

down the steps to the street. A light, two-wheeled *tonga* was parked at the kerb, a magnificent affair, a beautiful, high-stepping horse between the shafts, his brass harness gleaming in the lamplight.

The driver squatted on the pavement, chatting to an old beggar, but he sprang to his feet at once and ran forward. The Hindu desk clerk handed Janet in, gave the man his destination and then moved away.

The sky was scattered with the fire of a million stars, the moon so large that it seemed unreal like a paste-board cut-out. The wind blew in through the darkness carrying the last heat of the day across the river and she breathed deeply, wondering what the night might bring, her body shaking with a strange, nervous excitement.

The airstrip was half a mile outside Juma on a flat plain beside the river. It was not an official stopping place for any of the big airlines and had been constructed by the R.A.F. as an emergency strip during the war.

There was one prefabricated concrete hangar still painted in the camouflage of wartime, and the plane squatted inside, the scarlet and gold of its fuselage gleaming in the light of a hurricane lamp suspended from a beam.

Drummond leaned against a trestle table beside a wall-eyed Bengali merchant named Samil, Cheung's agent in Juma, and watched two porters load the narrow boxes into the plane.

'What's in this one?' he asked, kicking a wooden crate that carried the neatly stencilled legend *Machine Parts, F. Cheung, Esq., Sadar, Sikkim.*

Samil produced a bunch of keys, unfastened the padlock which secured the lid of the crate and opened it. He removed a mass of cotton waste and revealed a layer of rifles, each one still coated in grease from the factory.

Drummond took one out. It was a Garrand automatic, a beautiful weapon. He examined it closely and frowned. 'What about this?' He indicated the legend, *United States Army* on the butt plate. 'A bit stupid, isn't it? I don't think our American friends would be amused.'

'That's what they sent me this time,' Samil shrugged. 'Surplus stock always comes cheaper, you should know that.'

'Somehow, I don't think Cheung is going to like it.' He raised the Garrand, took an imaginary sight out of the door and stiffened suddenly as Janet Tate moved out of the shadows.

'What in the hell are you doing here?' he demanded.

'I'm sorry,' she said, her face serious. 'Hamid's gone off for the night. Before he left he called and told me you were out here. I thought you might like to take me to dinner or something.'

'Exactly what I intended to do.'

The two porters had stopped working and glanced at Samil uncertainly. Drummond was still holding the Garrand in both hands, close to his chest and Janet said gravely, 'Hamid said he thought you were loading motor spares.'

He put the rifle back with the others, wiped his hands clean on a lump of cotton waste and nodded to Samil. 'You finish up here. Nothing to worry about. I'll handle it.'

He turned, straightening his tie. 'How did you get here?'

'I came in a *tonga* from the hotel. I told the driver to wait.'

'Shall we go, then?'

He took her arm, aware of the stiff restraint, the tilt of her chin and knew that in some way he had disappointed her. In the *tonga* she sat silently in her corner, as far away from him as possible and Drummond chuckled.

'I'm sorry to spoil the image of the big bad gunrunner for you, but Ali knows damn well what I fly up to Sikkim in boxes labelled Machine Parts.'

She turned quickly in the darkness and he was aware of her perfume, delicate on the cool air.

'So does everyone else including the Khan himself.' He groped for her hand in the darkness and held it tight. 'Look, I'll tell you about it because it's coming to an end anyway and because I don't want my dinner spoiled. I've been looking forward to it.'

'Go on,' she said.

'There's a Chinese gentleman called Cheung up at Sadar. He's been there for six or seven months now. He's supposed to be a general merchant, but he happens to be an agent of the Chinese Nationalist Government on Formosa. He supplies the guns and I run them over the border into Tibet.'

'To help Tibetan guerrilla fighters against the Communist government?'

'Exactly.'

She reached over and touched his arm, the breath going out of her in a gentle sigh. 'Oh, Jack. I'm so glad.'

'Well that's a hell of a thing for a clean living Quaker girl to say,' he said. 'And don't go putting me on any pedestal. I do it for hard cash, not out of any political idealism.'

'You don't think the Tibetans stand any chance of winning then?'

He laughed harshly. 'Not in a thousand years. Their battle will be won or lost in other places. Vietnam, Malaysia, Sarawak, perhaps on the floor of the United Nations. But to hell with that. Where would you like to eat?'

'Somewhere full of colour, not just a tourist trap. I want to see the real India.'

'Good for you. We'll make a woman of you yet.'

They were moving into the centre of Juma now and he tapped the driver on the shoulder and told him to stop. 'We'll walk from here. You want to see the real India, I'll show it to you.'

He paid the driver, took her arm and they moved along the street. As they neared the centre, it became busier and busier. Vendors of cooked food squatted inside their wooden stalls beside charcoal fires, busy with their pans, the scent of spices and cooked meats pungent on the cooling air.

And then they turned into the old quarter where lamps hung from the houses and the bazaar was even more crowded than during the daylight hours as people walked abroad to savour the cool night air.

The pavements were jammed with wooden stalls, overflowing with masses of paper flowers, shoddy plastic sandals imported from Hong Kong, aluminium pots and pans looking somehow incongruous and out of place.

Craftsmen sat cross-legged in their booths behind the stalls of the brass merchants, still plying their ancient craft next to the silver-smiths and the garment-makers where they embroidered dancing girls' clothes.

There were Bohara carpets, rugs from Isfahan and, at the far end, prostitutes waiting in their booths, unveiled and heavily painted, and even here the curtain of night, the flickering lamps shining on cheap bangles and jewellery, cloaked the filth and disease, the squalor of the daylight hours.

They moved on, Drummond pushing to one side the numerous beggars who whined for alms, and finally turned into a narrow, quiet street leading to the river. Faintly on the night air, Janet could hear music. It grew louder and then they came to a narrow arched door.

'You wanted India? Well, this is it,' Drummond said.

They went along a narrow passage and came out on to a small landing at the head of a flight of steps overlooking a large, square room. It was crowded with Indians, mainly men, most of them wearing traditional dress. They were all eating hugely and talking loudly at the same time.

In the centre on a raised platform, a young, womanish *tabla* player, eyes rimmed with kohl, beat his drums with an insolent skill, looking around at the crowd as he did so, a bored and haughty expression on his face. His companion, an older man in baggy white trousers, three-quarter length black frock coat buttoned to the neck, looked strangely formal and played the *zita*, his fingers moving across the strings with incredible dexterity.

A small, neat Hindu in scarlet turban, his eyes flickering towards Janet with frank admiration, approached with a ready smile. 'A table, Mr Drummond? You wish to dine?'

'A booth, I think,' Drummond told him.

They threaded their way between the

tables, all eyes turning towards Janet and gasps of admiration, even clapping, followed them to their booth.

They sat facing each other across a small brass table, a bead curtain partially obscuring them from the other diners and Drummond ordered.

It was a simple meal, but superbly cooked. Curried chicken so strong that Janet gasped for breath, swallowing great draughts of cold water, thoughtfully provided by the proprietor, to cool her burning mouth. Afterwards, they had green mangoes soaked in syrup, followed by Yemeni *mocha*, the finest coffee in the world, in tiny, exquisite cups.

'Satisfied?' he asked her as he lit a cheroot.

She nodded, her eyes shining. 'Marvellous, I wouldn't have missed it for anything.'

'There's a floor show of sorts,' he said. 'Do you want to see it? Not exactly the Copacabana, I warn you.'

There was an unmistakable challenge in his voice and she responded immediately. 'I've never refused a dare since I was old enough to walk.'

'Suit yourself.'

There was a sudden roll on the drum, the lights dimmed a little and there was silence. There was an atmosphere of expectancy that she could sense at once and then a gentle, universal sigh echoed through the room.

A woman stepped through a curtain at the rear and poised for a moment, a dark silhouette against the light. 'Saida! Saida!' the name echoed faintly through the crowd.

'One of the few great *nautch* dancers left,' Drummond whispered to Janet. 'She's fifty if she's a day, but you'd never guess it.'

The right arm extended slowly and a tiny, tinkling cymbal sounded. Immediately the musicians responded on the *tabla* and *zita* and Saida started to sway sensuously, moving into the centre of the room.

Her face was heavily painted, a symbolic mask that never changed expression, but the body beneath the swirling, silken veils was that of a young and vibrant girl.

Gradually, the music increased in tempo and she moved in time, swaying from side to side, discarding her veils one by one until she stood before them, naked except for a small, beaded girdle low across her loins.

She stood quite still as the music stopped and the audience waited. The *tabla* player's fingers broke into a fast monotonous tattoo and she started to sway, hands above her head, clapping rhythmically, and the audience swayed with her, clapping in time, crying aloud with delight.

Round and round the perimeter of the floor she moved, faster and faster, sweat glistening on her body, until, with a sudden fierce gesture, she ripped the girdle from her loins and flung herself forward on her knees, sliding to a halt in front of a large, richly dressed merchant who squatted on cushions before a low table with two companions.

There was another abrupt silence and then the drum sounded again, slower this time, the beat becoming more insistent each moment as she writhed sinuously, thrusting her pointed breasts at him, twisting effortlessly from knees to buttocks, sliding away from his grasping hands, sharp cries rising from the crowd.

And then he had her, fingers hooking into her buttocks. As the crowd roared its approval, the drum stopped. She twisted from his grasp, her oiled body slipping between his hands, ran across the floor and melted through the curtain.

The musicians started to play again on a more muted key and the audience returned to their food, discussing the performance with much laughter and joking. When Drummond turned to look at Janet, her face was strangely pale.

'I warned you,' he said. 'You wanted to see the real India and this is a country where sex is as much a part of daily life as eating and drinking, an appetite to be satisfied, that's all.'

'Do you believe that?'

'Depends what a man's looking for, doesn't it? Had enough?'

She nodded and he called for the bill and paid it. The room was by this time heavy with smoke and there was the sound of drunken laughter everywhere. As they threaded their way between the tables, eyes turned on Janet, there were winks and leers and sly nudges.

Someone stood up at the edge of the floor and made an obscene gesture. There was a roar of spontaneous laughter and as she turned her head, flushing angrily, she was aware of a hand on her right leg, sliding up beneath the skirt.

She cried out in rage and mortification and swung round. There were four men seated at a low table, three of them typical of a breed to be found the world over in spite of their turbans and loose robes, young, vicious animals, spoiling for trouble. The man who had grabbed at her was older with wild, drunken eyes in a bearded face. He wore a black outer robe threaded with gold and his hands were a blaze of jewels.

As his chin tilted, the mouth wide with laughter, her hand caught him full across the face. His head rocked to one side, there was a general gasp and the room was silent.

His head turned slowly and there was rage and madness in the eyes. As he grabbed at her coat, Drummond spun her to one side.

The bearded man was only half-way to his feet when Drummond's right foot swung into his crutch. The man screamed, doubling over, and Drummond raised a knee into the descending face, smashing the nose, sending him crashing back across the coffee table.

And the thing Janet couldn't understand was the silence. No one moved to stop them when Drummond turned, straightening his jacket, took her arm, and pushed her through the crowd to the stairs.

Outside in the street, he urged her on, turning and twisting through several alleys until, finally, they emerged on an old stone embankment above the river.

'Why the rush?' she said. 'Did you think they might follow us?'

'That's the general idea.' He lit a cheroot, the match flaring in his cupped hands to reveal the strong, sardonic face. 'The young squirt-about-town I treated so roughly back there happens to be the son of the town governor.'

'Will there be trouble?'

'Not the official kind, if that's what you

mean. He's got away with too much in the past for anyone to start crying over his ruined looks at this stage. He might put someone on to me privately, but I can handle that.'

'Did you really need to be so rough?'

'It never pays to do things by halves, not here. This isn't tourist India, you know. The only thing I'm sorry about is taking you there in the first place. I should have had more sense.'

'I'm not,' she said. 'You weren't responsible for what happened. To tell you the truth, I rather enjoyed myself.'

'Including the *nautch* dance?'

She laughed. 'I'll reserve my opinion on that part of the programme. It was very educational, mind you.'

'Something of an understatement. You know, you're quite a girl, and for someone who believes in turning the other cheek, you throw a good punch. You certainly rocked him back there.'

'A quick temper was always my besetting sin,' she said. 'My old grannie used to warn me about that when I was a little girl back home in Maine. Quakers are really quite nice when you get to know them. Flesh and blood, too.'

He grinned and took her arm. 'All right, I surrender. Let's walk.'

They went on to the beach below the embankment and strolled through the moonlight without talking for a while. Now and then, sandbanks collapsed into the water with a thunderous roar and cranes threshed through the shallows, disturbed by the noise.

Huge pale flowers swam out of the night, and beyond the trees the sky was violet and purple, more beautiful than anything she had ever seen before. They passed a solitary fisherman cooking a supper of fish over a small fire of dried cowdung and Drummond gave him a greeting in Urdu.

'What do you do in Balpur beside fly in guns for Mr Cheung?' she said after a while.

'Survey work for the Indian government, freight general cargo or passengers. Anything that comes to hand.'

'I shouldn't have thought there was much of a living in that.'

'There isn't but Cheung pays well for the Tibetan trips. And I'll be leaving soon, anyway. I've had enough of the place.'

'What's it like?'

'Balpur?' he shrugged. 'Barren, treacherous mountains. A capital of three thousand people that's more like an overgrown village. An army, if you can call it that, of seventy-five. When winter comes, it's absolute hell and that's in another month. The roads are the worst in the world at the best of times, but during the winter, they're completely snowed up.'

'What about the Khan?'

'An old mountain hawk, proud as Lucifer. Quite a warrior in his day. To his people, something very special. Not only king, but priest, and that makes for quite a distinction. You'll like Kerim, his son. A great pity about his accident. I hope your people in Chicago can fix him up all right.'

'He's eight, isn't he?'

'Nine in three months.'

'My instructions told me to get in touch with a Father Kerrigan when I arrive. Apparently he's in charge of all the arrangements.'

'You'll like him,' Drummond said. 'He's about sixty. A marvellous old Irishman who just won't give in. He's been twelve years in Sikkim and hasn't made a single convert and the people adore him. It's fantastic.'

'If he hasn't got a congregation, what does he do with himself?'

'As it happens, he's a qualified doctor. Runs a small mission hospital about a mile outside of Sadar, completely on his own. There's one other European up there, a man called Brackenhurst. A geologist for some British firm or other. They've also made him British Consul, but don't let that impress you. It doesn't mean a thing.'

'You don't like him, I take it?'

'Not much.'

He stopped to light another cheroot and she said casually, 'Why did you leave the Navy, Jack?'

He paused, the match flaring in his fingers, his eyes dark shadows. 'You really want to know?'

She didn't answer and he shrugged, flicking the match into the night. 'They kicked me out, or advised me to leave, which comes to the same thing for a career officer.'

She could sense the pain in his voice and put a hand on his arm instinctively. 'What happened?'

'I was a Fleet Air Arm pilot during the Korean War. One bright morning in July, 1952, I took my squadron to the wrong target. When we left, it was a smoking ruin. We did a good job. We managed to kill twenty-three American marines and ten Royal Marine Commandos who had been serving with them.'

There was bewilderment in her voice. 'But how could such a thing happen?'

'The briefing officer gave me the wrong information.'

'So it wasn't your fault?'

'Depends how you look at it. If I'd checked my orders more carefully, I'd have spotted the mistake. I was too tired, that was the trouble. Overtired. Too many missions, not enough sleep. I should have grounded myself weeks before, but I didn't.'

'So they couldn't court-martial you?'

'A quiet chat with someone with gold rings

all the way up to his elbow, that's all it took. I got the message.'

'I'm sorry, Jack. Sorrier than I can say.'

Her voice was warm and full of sympathy. They had reached a flight of stone steps leading up from the shore and he paused and looked at her.

Her mouth opened to cry a warning and he ducked, turning to meet the rush of feet from the darkness.

A fist grazed his cheek, he lost his balance and rolled over and over, hands protecting his genitals as feet swung in viciously.

He sprang up and backed to the wall. There were three of them, dark, shadowy figures in tattered robes, scum from the market place hired for a few rupees. Above them on the steps below the lamp, stood the man from the cafe, supported by two of his friends, blood on his face.

A knife gleamed dully and Janet ran in past the three men to join Drummond against the wall. 'Kill him!' the bearded man cried. 'Kill the swine!'

Drummond was tired. It had been a long

evening. His hand disappeared inside his coat reaching to the leather holster on his left hip and reappeared holding a Smith & Wesson .38 Magnum revolver with a three-inch barrel.

He fired into the air and there was a sudden stillness. 'Go on, get out of it!' he shouted angrily and fired a shot towards the man on the steps that ricocheted into the night.

The men from the market place were already running away along the shore, cursing volubly, and the governor's son and his two friends staggered into the darkness.

Drummond slipped the revolver back into its holster and looked down at her calmly. 'You know, I really think it's time we went back to the hotel, don't you?'

She started to tremble uncontrollably and he reached out, pulling her into his arms. 'It's all right. Everything's all right now.'

He stroked her hair gently with one hand and his lips brushed her forehead. In the heavy stillness of the night, she could almost hear her heart beating. When he tilted her chin and kissed her gently on the mouth, it was like nothing she had ever known before.

He slipped her arm in his without speaking, and together they went up the steps to the embankment.

4

The Last Place God Made

The air was bumpy as they flew out of the pass for a forty-knot wind was blowing across the mountains. They climbed through a heat haze that was already blurring the horizon and levelled out at 9,000 feet to cross the mountains between India and Balpur.

Janet Tate was in the front passenger seat beside Drummond and Hamid sat behind her. She was wearing a white blouse, collar turned down over the neck of a cashmere sweater, cream whipcord slacks and a sheepskin coat that Drummond had provided.

Hamid poured coffee into a plastic cup and handed it to her. 'We're moving into Balpur now,' he said. 'The mountains to the east are in Bhutan with Assam far beyond in the haze. The Chinese broke through in strength there in 1962.'

'Were you there?'

He shook his head. 'No, I was on the Ladakh front in the north-west.'

'It was supposed to be pretty bad up there, wasn't it?'

'A vision from hell,' he said grimly. 'Can you imagine what it's like trying to live at 20,000 feet, never mind fight? The mules died of asthma, the men of pulmonary oedema. You've heard of it, I suppose.'

She nodded. 'The lungs fill with water, don't they?'

'An ironic way for a man to die in battle – by drowning. We could never get them down to the base hospitals in time for treatment, that was the trouble.'

'Hadn't you any air support, helicopters?'

He laughed harshly. 'Until October, 1962, we hadn't needed them. The way of peace was the way for India.' He shook his head. 'No, we didn't have the necessary planes. Even if we had, there weren't the pilots.

Certainly not the kind who could fly in that sort of country. That's where I met Jack, you know.'

She turned to Drummond in surprise. 'You were flying for the Indian Army?'

'Five hundred quid a week,' he said. 'Good money by any standards.'

'Don't listen to him,' Hamid broke in. 'A game he plays. From Leh, he flew three operational flights a day into the Ladakh mountains to one small airstrip at 18,000 feet, taking in supplies and ammunition, bringing out the sick and wounded. In five weeks, he flew just over a hundred sorties, then collapsed and spent three weeks in hospital suffering from complete exhaustion. His contract called for five flights a week, no more.'

'He should have added that they didn't pay me for the time in hospital,' Drummond told her. 'That's the wily oriental for you.'

He increased speed and banked in a long, sweeping curve that took them out of a shallow pass and into a valley beyond. A broad river flowed sluggishly, snaking between jagged cliffs, a thread of silver in a landscape so savage and sterile that it took the breath away.

'Remember what I told you,' Drummond said. 'The last place God made. And to think the Chinese have laid claim to this bloody lot.'

'But why?' she said.

'The same psychology the Roman Emperors used,' Hamid told her. 'Give the mob circuses to take their minds off the more important problems. In China in 1962, the harvest was bad and thousands starved, so their army invaded India, a country completely unprepared for such an attack, and presented their people with a ready made victory. In Pekin, they were able to tighten their belts and wave banners.'

'Have they really laid claim to Balpur?'

'Along with almost every other border country. Actually, Balpur *was* a part of the Chinese Empire in ancient times. The people are Mongolian. Only the ruling class are Muslims, descendants of the original invaders. But no one seriously imagines that they would invade. For one thing, the old Khan has preferred to stay completely neutral. He's the only ruler of a border state who hasn't signed a mutual defence pact with India.'

'Yet he accepts you as an adviser?'

'To an army of seventy-five men. A political gesture only. In Pekin they laugh about it.'

She almost mentioned Mr Cheung, but remembering what Drummond had told her on the previous evening, kept silent. Even if Hamid did know the truth, that Cheung was in fact a Chinese Nationalist agent, that Drummond was flying in guns to Tibetan guerrilla fighters, he would probably prefer to know nothing officially. Remembering Vietnam, she sighed heavily. The same pattern, violence, blood and suffering turning on each other in a circle that had no ending.

They were flying at no more than a thousand feet above the floor of the valley and suddenly, in a bend of the river, she saw Sadar, flat-roofed houses scattered untidily across a broad plateau, the Khan's palace like a fortress in a walled garden.

The Beaver banked tightly and swept in past the graceful tower of a mosque, and beyond the town on the plain to the south she could see the airstrip, a narrow slot laboriously carved out of the rough terrain, a windsock on a tall pole at one end. Drummond circled once then turned into the wind for a perfect landing between two rows of empty oil drums.

There was a small improvised hanger constructed of rusting corrugated iron, barely large enough to house the Beaver from the look of it. He taxied towards it and switched off the engine.

He unfastened his seat belt, jumped to the ground and turned to give Janet a hand. At the same moment, a Land Rover appeared from among the houses on the edge of the town and came towards them in a cloud of dust.

She shivered and wrapped her sheepskin coat more tightly around her. 'It's colder than I thought it would be.'

'Winter coming,' Drummond said. 'Maybe it'll be early this year.'

An old army jeep, still painted in the grey-green camouflage of wartime, its canvas tilt patched and mended in many places, was parked inside the hangar. He and Hamid had just started to transfer the luggage to it from the plane when the Land Rover arrived.

Mr Cheung jumped out of the passenger seat and came towards them wearing a heavy blue quilted jacket and an astrakhan hat. His driver was a young fair-haired man with a bronzed, reckless face. He wore a sheepskin jacket in untanned hide and knee-length boots. A revolver, slung low on his right hip in a black holster, seemed theatrical and out of place.

He came forward with a ready smile, eyes fixed on Janet, and Hamid said maliciously, 'Why the gun, Tony? Expecting trouble?'

The young man flushed. 'I'm driving up to my base camp at Howeel for a couple of days. They'd cut your throat for the shoes on your feet up there. I've come for that new theodolite I ordered if Drummond's remembered to bring it.'

'It's in the plane,' Drummond said coldly. 'Help yourself.'

'So this is Miss Tate?' Cheung took both of her hands in his. 'We must try to make your stay a pleasant one.'

'You knew I was coming?'

Hamid grinned. 'I had Indian Army Headquarters in Juma send a signal to warn the Khan.'

Cheung nodded. 'Colonel Dil got the message last night by radio.'

'And probably told you before the Khan.'

Brackenhurst jumped down from the Beaver and turned to lift out a wooden case containing his theodolite. 'A hell of a lot of machine parts you seem to bring through these days,' he commented and turned to Janet before Drummond could reply. 'I'm Tony Brackenhurst, Miss Tate. I'm doing geological survey work up here, but I'm also the British Consul. If I can help you in any way, don't hesitate to ask.'

'She happens to be an American, so that's hardly likely,' Drummond said acidly.

Brackenhurst ignored him, holding her

hand for longer than was necessary, an eager smile on his lips, and it was the smile which betrayed him, somehow revealing an essential weakness, a lack of strength.

'Why, that's very kind of you, Mr Brackenhurst.'

'I'll be back in two days,' he said. 'You'll probably still be here from what they tell me of the boy's condition.'

He carried the theodolite across to the Land Rover and Cheung said quickly, 'I'll go back with him. You'll have enough in the jeep with the three of you and the luggage. You'll call on me this afternoon, Jack?'

'After lunch. I'll take Janet out to the mission first. Is the boy still out there?'

Cheung nodded and smiled down at her. 'And you, I will have the pleasure of seeing you again this evening, Miss Tate. The Khan is to give a small dinner party for you. He has honoured me with an invitation.'

'I'll look forward to that, Mr Cheung.'

The Land Rover moved back towards town and Drummond drove the jeep out of the hangar. He and Hamid pushed the Beaver inside and padlocked the door.

'I'll take Janet out to Father Kerrigan now. What about you, Ali?'

Hamid shrugged. 'You can drop me at Colonel Dil's headquarters. I'll probably see you both tonight at the palace unless the old boy's decided to change his usual guest list.'

They got into the jeep and Drummond drove towards the town, following the rutted track that did service as a road. He changed down, scattering a herd of goats, and they entered the outskirts of Sadar.

Janet looked about her with interest, but there was nothing of the gaiety and colour of Juma and Altaf here. The people were small, squat Mongolians with skins the colour of weathered parchment and slanting eyes. The men wore boots of untanned hide, baggy trousers and sheepskin jackets. Only a few sported the turban, the majority preferring conical sheepskin caps with earflaps. The women's attire differed in only one significant detail. Instead of the sheepskin jacket, they wore three-quarter length blanket coats of black and brown, relieved in some cases by a necklace of silver coins.

They were dour and unsmiling, drab as the rocky land that bred them. Even the children in the market place lacked the energy and humour of their Indian counterparts, and there was a strange absence of bustle and vitality as they drove through the bazaar.

'No one seems to smile,' Janet said. 'Have you noticed that?'

'This is a poor country,' Hamid told her. 'Anything they get has to be squeezed out of the very rocks. Life is hard, work from dawn till dusk. It leaves little time for laughter.'

Across the square stood a barrack-like building, the flag of Balpur, a black eagle against a grey and gold background, lifting in the slight breeze above the entrance. Two sentries, almost incongruously smart in neat khaki uniforms and military turbans, presented arms as Drummond braked and Hamid got out.

He reached for his canvas grip and an orderly ran down the steps and relieved him of it. 'I'll see you tonight, then,' he said and his hand lifted in a brief salute.

The palace was a hundred yards further on and looked considerably less forbidding than it had done from the air, wrought iron gates standing open to reveal a gravel drive, tall cypress trees fringing the wall, a profusion of greenery beyond to where a fountain lifted gracefully into the calm air.

'I must say that looks rather more inviting,' Janet remarked.

'Not surprising,' Drummond said. 'The Khan's a Muslim, remember. At least they know how to live.'

'What's the religion of his people generally?'

'A lot pay lip service to Islam and a great many still adhere to Buddhism, but in a bastardised form. And then there's a minority group of Hindus who've kept themselves apart over the centuries. Not more than two or three thousand in the entire country.'

They were by now moving out of the town again and the houses were more scattered, two-storeyed walled villas in the main, obviously the homes of the rich of Sadar, whoever they were.

Drummond slowed, swung the jeep in through an arched entrance and braked to a halt in the courtyard of a small bungalow surrounded by a walled garden.

'This is my place,' he said. 'If you don't mind hanging on, I'll drop my things and be straight out again.'

As he got out, a small, greying woman, swathed in a dark robe, her face seamed and wrinkled, opened the front door and moved out on to the verandah inclining her head in greeting, hands together, Indian style.

'Your housekeeper?' Janet asked.

He nodded and reached for his canvas holdall. 'I won't be a minute.'

'Mind if I come in?' she said. 'I'd love to see inside.'

He hesitated perceptibly and then shrugged. 'If you'd like to, but there really isn't much to see.'

She followed him up the steps. At the top, he murmured something quickly to the old woman who went back in, then stood to one side. 'After you.'

She found herself in a narrow entrance hall with rough cast walls and a floor of polished wood. He opened a door to the right and she moved into the main living room. There was a great stone fireplace, skin rugs on the wood floor and the furniture was of the simplest; a dining table, several easy chairs and a couple of shelves of books.

'I'll be with you in a minute,' Drummond said and he crossed the room and went through another door.

She walked slowly around the room, examining everything and paused at the bookshelves. There was a small figurine of a dancer on the table beneath, carved from some dark wood of incredible hardness. She picked it up and examined it closely. The breasts were of a ripeness that was almost lifelike, hands extended in a ritualistic pose, the unsmiling, grave face fixed for all eternity. There was a slight sound from behind and she swung round and found a woman standing in the doorway to the hall.

Like the old housekeeper she was an Indian, but quite young with a pale, flawless complexion, set off to perfection by her scarlet sari. There was a silver rope necklace around her neck, gold bracelets on the wrists and her dark eyes were rimmed with kohl.

In that same moment, Drummond came in from the bedroom. He said something quietly in Urdu and the girl turned at once and disappeared into the hall.

'Who was that?' Janet said.

'The old girl's daughter, Famia.' He took the figurine gently from her hands. 'You like this?'

'Yes, is it very old?' she replied automatically.

'Greco-Buddhist. Probably second century. You'll find things like this all over Balpur. As I said before, Buddhism used to be very strong up here, real Buddhism, I mean. Monasteries all over the place.'

'Are there any left?'

'One or two.' He glanced at his watch. 'We'd better get moving. It's almost eleven o'clock and Father Kerrigan holds his daily surgery at half-past. We'll try and catch him before it starts.'

They went out to the jeep and he handed her in and drove away as if nothing had happened. But things were not the same and there was a constraint between them that had not been present before.

Janet remembered the girl, her shapely body, the pale beauty of her skin against the scarlet sari, and a burning anger took possession of her that she found impossible to analyse.

The mission was on a hill above the river. It was a long, low, flat-roofed building, walled in by grey stone, as seemed to be the custom with all houses in this stark country, and the tiny belfry of a small chapel reared above it.

Flocks of goats, sheep and a few small horses grazed on the sparse grass at the entrance, and thirty or forty people waited patiently, squatting on the ground or leaning against the wall.

As Drummond slowed the jeep to drive

through, Janet leaned out, her trained eye quickly taking in the evidence of disease. Rickets and ringworm in the children, old people with faces eaten away by yaws, eyes encrusted with dried pus and, here and there, a broken limb held awkwardly in a crude bandage.

'He doesn't handle all this on his own?' she demanded, turning to Drummond as they drove in through the entrance and braked at the bottom of a flight of stone steps.

He switched off the engine and nodded. 'Don't ask me how, but he does. Has an old woman to do the cooking, but that's all. Here she comes now.'

The woman who opened the front door and came out into the porch had the same ageless Mongolian face as the people in the market place, but wore a long cotton skirt and an Indian Army issue khaki sweater with cloth epaulets. The red scarf around her head and gold ear-rings made her look like a gypsy.

Drummond went up the steps with Janet's two cases, put them down and spoke to her in slow, careful English. He came back down the steps and took Janet's arm.

'He's in the chapel.'

They crossed the courtyard to the tiny, grey-stone building, he opened the heavy wooden door and they went inside. The lights were very dim, and down by the altar the candles flickered and the statue of the Holy Mother seemed to float out of the darkness.

Father Terence Kerrigan knelt in prayer, his rugged, stubborn old Irish face momentarily relaxed, almost childlike in its purity, his white hair gleaming like silver. When he crossed himself and got to his feet, she saw that he was a big man, built like a tree with shoulders as wide as Hamid's.

He turned, narrowing his eyes short-sightedly when he saw them there in the shadows and came forward with a ready smile.

'Jack, is it yourself, and this will be Miss Tate?' He took her hands in his, holding them tightly. 'It's good to see you here, my dear. I got word from Colonel Dil that you were coming in today. He had a message last night from Ali Hamid over the radio.'

'I feel like a fraud, Father,' she said. 'I believe you were expecting a doctor.'

'Nonsense, my dear, a qualified nursing sister with two years' experience in Vietnam refugee camps will do for me any day of the week.' He chuckled at her astonishment, 'Major Hamid is always most thorough.'

They crossed the courtyard, mounted the steps and went inside. The entrance hall had been turned into a dispensary, the stone walls whitewashed, drugs, medicines and equipment neatly arranged on white-painted shelves giving an overall impression of cleanliness and efficiency.

'This is where most of the work is done and as I'm the only qualified doctor in Balpur the pace is usually fast and furious.' He glanced at his watch. 'You'll see for yourself in precisely fifteen minutes.'

'What about my patient?' Janet asked.

'Kerim?' the old man sighed. 'Frankly, he's not been too marvellous. He's been staying here, of course, so that I can give him constant supervision. The Khan wanted me to take up residence at the palace, but naturally, I had to refuse. As I pointed out, I do have other patients.'

'And how is Kerim now?'

'Rather better. He's been very feverish, but we seen to be over the worst of that now. In any event, I think we should wait for a few days before contemplating such a long journey.'

'So Janet stays here?' Drummond said.

'If she can put up with a crotchety old fool.' Father Kerrigan smiled. 'Would you like to take a peep at Kerim?'

He led the way through into a narrow whitewashed passage and opened a door on the left. The boy looked very frail as he slept, head turned to one side on the white pillow, a heavy bandage crossing his right eye and they withdrew softly.

The priest opened the opposite door and ushered her into a small room, simply furnished with a narrow bed and wooden locker. The one touch of luxury was a large sheepskin rug on the floor. A french window opened on to a verandah overlooking an overgrown and neglected garden.

'The best I can do, I'm afraid,' he said apologetically.

'A palace compared to what I was used to in Vietnam.'

They returned to the dispensary and found Drummond standing at the door looking outside. The courtyard had filled with people, all squatting together in the dust, waiting patiently for the old priest to begin.

He took out his watch again and pursed his lips. 'Five minutes late. This will never do. I'll have to say goodbye for now, Jack. We'll be seeing you tonight at the palace, I imagine.'

'I expect so.'

Drummond turned to Janet, but she was touching Father Kerrigan on the arm as he moved away. 'Could I help, Father?'

The old man looked down at her searchingly and then a slow smile broke across his face. 'I'd be glad to have you, my dear. I'll find you a robe.'

She nodded briefly to Drummond. 'See you tonight, Jack.'

She turned away, different now, holding

herself straighter, competent, assured. She and the old man stood at the back of the dispensary, talking as she pulled on the white robe he had found for her, a strange intimacy between them.

Drummond turned abruptly, pushed his way through the crowd, climbed into the jeep and drove quickly away.

5

Dinner at the Palace

Through the french windows, the white balustrade of the terrace shimmered palely and the tall cypress trees were silhouetted against the evening sky. From the garden came the timeless, incessant chirping of the crickets.

Inside, the soft lamplight gleamed on delicate crystal decanters and silver and gold tableware, and the great ruby in the centre of the Khan's turban glowed dimly like an ember stirred by a soft wind.

He was seventy years of age, but carried himself well in his London tailored mohair and silk dinner jacket, and the face beneath the turban was still that of a warrior, proud and strong with the touch of arrogance of one born to rule.

He sat at the head of the table, Janet Tate on his left, and he turned to her with a smile, speaking in careful, precise English. 'More brandy, Miss Tate?'

'I don't think so, thank you.'

'A little more coffee, then?'

He snapped a finger and a servant came forward quickly. There were five of them at the table besides the Khan, Janet, supremely beautiful in her simple black silk dress, and Jack Drummond on her left in a white dinner jacket. Father Kerrigan sat on the Khan's right hand next to Mr Cheung, and Hamid and Colonel Sher Dil, commander of the Khan's small army, faced each other, magnificent in dress uniform.

'Father Kerrigan has made you comfortable, Miss Tate?' the Khan asked.

'He couldn't have done more.'

The Khan sighed. 'It would have pleased me to have had you as my guest here at the palace, but he is a stubborn old man.'

'And if that's true, then I know another

not a thousand miles from here,' the priest said, speaking with the familiarity of an old friend and reaching for the brandy decanter. 'Would you imagine it, Janet, he wanted me to forsake every other blessed patient I have, close the mission and move in here?'

The Khan shrugged helplessly. 'What can one do? He even refused the soldiers I sent. At this moment, who guards the Hope of Balpur?' he challenged the old priest.

'Tell me first who in Balpur would harm him,' Father Kerrigan countered.

The Khan sighed. 'You see, Miss Tate, I am not even ruler in my own house.'

'If you must know, old Nerida's sitting at the boy's bedside this very moment,' Father Kerrigan told him. 'She'd cut off her arm rather than move from that spot before I return.'

'You have seen Kerim today?' the Khan said to Janet. 'He is well?'

She nodded. 'But still a little weak. An injury of this kind is a great shock to the whole system, especially for a child.'

'A child who will be a man in another

three years, an important distinction. Under our customs, he must then be presented to the people, ready to take my place if need be. That is why I am anxious that he starts on the journey to America with as little delay as possible.'

'We must wait for another few days,' Father Kerrigan said. 'I'm sure Miss Tate agrees with me.'

The Khan glanced at Janet and she nodded. 'I think Father Kerrigan is right. And we've time to spare. Kerim can be back within a month of the operation, you know.'

He threw his arms wide. 'Then I must bow before the wind. You play chess, Miss Tate?'

'Not very well, I'm afraid.'

'Father Kerrigan considers himself a master. It is my painful and frequent duty to prove otherwise.'

'Indeed, so?' the priest said, pushing back his chair and rising, glass of brandy in hand. 'If your Highness would be good enough to lead the way to the usual place, we can get down to the business of making you eat your words.'

'A pleasure.' The Khan got to his feet and looked enquiringly at the others. 'Gentlemen?'

Hamid glanced at Drummond and Sher Dil. 'Billiards?'

They both nodded and Cheung smiled across at Janet. 'Which leaves Miss Tate and myself. With the Khan's permission, perhaps I could show her some of the treasures of the palace?'

'Please do. It should take me no longer than an hour to encompass the downfall of this turbulent priest.'

'Is that a fact, now?' Father Kerrigan said in mock anger and they went out.

Hamid, Sher Dil and Drummond had their heads together for a moment, something to do with a report over the radio from Indian Army Headquarters about patrol clashes in the Ladakh area. Cheung joined them and Janet moved to the window and looked into the garden.

It was very beautiful. Great, Grecian-style jars were spaced along the terrace, filled with dwarf iris, and the scent of hibiscus was heavy on the night air. Lower down in the shadows, the slender cypress trees stood like straight sentinels, dark against the sky, and the moon was full.

Cheung paused beside her. 'A startling contrast, isn't it? In here all the beauty in the world, a garden by night. Beyond those walls, a harsh, sterile land where even mere existence is a struggle.'

'Has it always been this way?'

He nodded. 'In the old days, the tribesmen raided into India like wolves. Their name was a byword for cruelty. But those days are gone. Now they must live off the land and the land has little to give.'

'Can nothing be done?'

He shrugged. 'Who knows? Brackenhurst may turn up with something in his survey, evidence of mineral deposits worth developing, perhaps. The Khan has his hopes, but I doubt if they will come to much. Brackenhurst would not be the first geologist to waste his time here.'

'And yet Hamid tells me the Chinese Government in Pekin has laid claim to Balpur.'

'And Nepal and Bhutan, even parts of Assam.' He shrugged. 'Words, merely words. But as a matter of interest, there can be little doubt that in other times Balpur *was* part of the Chinese Empire. Come, I will show you.'

They went back inside, moved into the central hall, and he opened another door. The room was in darkness. Janet heard the click of the switch, but was totally unprepared for what followed.

On every side a row of glass showcases, each with its own illumination, sprang into view to float in the darkness. But it was their contents which drew from her an involuntary gasp of admiration. They contained the most superb collection of pottery she had ever seen.

There were alabaster jars, pale, translucent and delicate, glazed urns in red and black, their colours as vivid as on the day they had been fired.

Most were unmistakably Chinese and others showed a distinct Chinese influence. There was also a collection of figurines like the one she had seen at Drummond's bungalow.

'Jack has one of these,' she said. 'He told me it was Greco-Buddhist.'

'That's right. As you're probably aware, Alexander the Great invaded India. Amazing to what extent Greek culture penetrated the entire border area and yet in India, their literature doesn't even mention his name.'

She reached out and touched a delicate and beautiful wine jar which had been painstakingly put together piece-by-piece to judge by the network of fine lines that covered it.

'Where was this discovered?'

'A burial mound south of the city near the river. There are many such sites. You must visit some of them while you are here. There is a most interesting ruin of a Buddhist temple not far from the mission. Breathtaking by moonlight. I can recommend it.'

He smiled charmingly and Janet, hesitating, was saved by Hamid who entered at that moment. 'There you are.'

'I thought you were playing billiards?' she said.

'A variation of our own, a sort of knock-out competition. Jack and Sher Dil were too good for me.'

'Mr Cheung was showing me the Khan's collection. I had no idea the Chinese had so much influence in this area in ancient times.'

'God knows why,' he said. 'The damned place must have been an economic liability even in those days.'

Cheung glanced at his watch. 'It's getting late. I really think I must be going. I leave Miss Tate in your capable hands, Major.'

He went out quickly and she turned to Hamid with a sigh. 'I feel rather sorry for him. We've been standing here talking about the splendours of China past and he isn't even permitted to be a part of China present. It must be a terrible thing to be an exile from one's own country.'

'The tragedy of the twentieth century,' Hamid said. 'Did you like him?'

'It's difficult to say. He puts himself out to be pleasant, but I feel that he's looking at me from behind his eyes if you know what I mean.'

'An excellent description. What about Sher Dil?'

'A wonderful man. He's so beautifully correct, so . . .'

She hesitated and Hamid chuckled. 'So positively British? The Imperial taint still lies heavy upon us. Sher Dil was a Sword of Honour man at Sandhurst in his day. He was also a colonel in the Indian Army as long ago as 1945.'

'What went wrong?'

'For many years we tried to follow the way of peace in India. Nehru was sure that such a neutrality would be respected by all. Many men like Sher Dil, high ranking regular army officers, were not so sure and said so. When the army was reduced, they were the first to go.'

'And so Sher Dil came here?'

'To command an army of seventy-five men for the Khan, most of them recruited in India. The locals don't take kindly to uniforms.' He laughed. 'But a night like this is made for love and laughter and nothing else. I will show you the delights of the garden.'

'Which Mr Cheung has already done.'

'Not with my superb efficiency.'

They left the display room and moved out to the terrace, pausing at the top of a flight of shallow steps for Hamid to light a cigar.

The moon was caught in the dark meshes of the cypress trees, the night air heavy with the scent of flowers and a fountain splashed into a fish pool amongst the trees as they went down the steps, her hand on his arm.

'The Hour of the Dove they call it.' He waved a hand theatrically. 'The time for lovers to unburden their hearts to each other.'

They came to the fountain in the centre of the garden and she sat on the low wall that ringed the pool, dabbling her hand in the water, and somewhere a bird called sweetly through the night.

'This place is like finding the Garden of Eden in the wilderness. How does he do it?'

'An army of gardeners and careful cultivation and the walls keep out the winds, remember.' Hamid breathed in deeply and sighed. 'And the strange thing is that it can all die in a single night. When winter comes here, it strikes suddenly, like a sword biting into warm flesh.'

She gazed down into the moonlit water, watching the fish nibbling gently at her trailing fingers. 'Jack told me about what happened in Korea.'

Hamid raised her chin with one hand and looked into her eyes. 'You like him, don't you?'

'Very much. I've never met anyone quite like him. He's a strange man, violent and bitter, and yet he can be the gentlest person I've ever known.'

'The story of my life.' Hamid sighed. 'What would you like to know?'

'We stopped at his bungalow this afternoon. There was a girl there. Famia, I think he called her.'

'His housekeeper's daughter.'

She seemed to hesitate and then plunged on, 'Is she his mistress?'

'So that's it?' Hamid chuckled gently and took her hands. 'He's a grown man, Janet, not a boy. There would be something strange if he didn't feel the need for a woman occasionally, now wouldn't there?'

Momentarily, her hands tightened on his as anger swept through her like an uncontrollable fire, and Hamid touched her gently on the right cheek.

'Poor Janet. India makes a harsh taskmaster.'

'I think I love him, you see,' she said in a low voice. 'It's as simple as that.'

'It's never as simple as that,' he said solemnly and pulled her to her feet. 'I think we'd better go back while I can still remember that fact.'

'Just one more thing,' she said. 'Is he as embittered over this Korean business as he appears to be sometimes?'

Hamid shook his head. 'Not really. He's too intelligent to blame himself for what was really an accident of war, but he loved the Navy. That was his greatest loss.'

'And what does he believe in now?'

'Nothing. At least this is what he tells himself, and spends his time living dangerously, working for the highest bidder to amass a fortune.' He chuckled gently. 'Only to end by betraying all his hard won principles when he looks upon the face of suffering, as he did in Ladakh during the Chinese invasion.'

'You like him a great deal, don't you?'

'I value real friendship,' he said simply. 'Jack Drummond has shown that to me many times.'

They walked back through the garden in silence. As they mounted the steps to the terrace, Drummond came through the windows.

'There you are. Father Kerrigan thinks he should be going. He doesn't like leaving Kerim for too long. I'll run you back in the jeep.'

'I'll get your coat,' Hamid told her and went inside.

'Did you win your game?' she asked.

'No, did you?'

Janet smiled faintly. 'You couldn't be more wrong.'

She brushed past him and went inside and Drummond stood there in the half-darkness, listening to the rise and fall of the voices, a cold finger of excitement moving inside him, leaving his stomach hollow and empty.

She sat next to him on the way back, Father Kerrigan on the other side, and now and then the wind lifted the edge of her silken head-scarf into Drummond's face.

He was aware of her warmth, the softness of the thigh against his, the delicate perfume, and gripped the wheel tightly, inhaling her sweetness, aware of feelings he had not experienced for a very long time.

The old priest kept chuckling to himself. 'I wish you could have been there, the pair of you. And didn't I show him? It'll be many a long day before he comes crowing over Terence Kerrigan again.'

Drummond glanced at Janet and grinned as he turned the jeep into the courtyard of the mission. 'I think he must have won.'

'Ah, get away with you!' The old man snorted as he got out of the jeep and then smiled, his face clear in the moonlight. 'A fine night for a drive.'

Drummond hesitated and Janet said calmly, 'Mr Cheung mentioned the ruins of a Buddhist temple not far from here. He seemed to think they were worth seeing by moonlight.'

'And maybe he had a point there.' Father Kerrigan slapped the side of the jeep with his bare palm. 'Off with you, now, and don't be late.'

Drummond took the jeep out through the entrance and turned across the moonlit plain beside the river. He had taken down the canvas tilt earlier and the wind was sharp and cold, carrying with it the scent of wet earth. A few minutes later, they came over the edge of an escarpment and the ruins of the temple lay before them in the centre of a small plateau, bare and windswept, crumbling with the years.

He braked, switched off the engine and they walked the last few yards. The full moon touched the scene with a pale luminosity and the dark shadows of half-ruined pillars fell across the mosaic floor like iron bars.

The statue of the Buddha was at the far end, chipped and cracked by time and the weather, one arm missing, but the great, serene face was still complete, hooded eyes staring blindly into eternity across the river.

Janet walked towards it slowly and Drummond paused to light a cheroot. When he raised his head, she was standing at the edge of the crumbling terrace, staring pensively into the night.

The moon was directly behind her, outlining her shapely limbs through the thin silk of the dress, and when she turned and looked at him, she looked unreal and ethereal like some dark goddess of the night who might fly away at any moment.

They stood like that, trapped by a moment of time, looking at each other, and then she came forward slowly, reached up and gently touched his face.

Drummond turned his head, brushing her palm with his lips and slipped his arm about her waist. She leaned against him, trembling a little and in the distance, thunder rumbled menacingly.

She glanced up quickly. 'What was that?'

'Storm on the way.' He pointed to where sheet lightning flickered over the mountains. 'We'd better get moving.'

She was conscious of the unnatural stillness.

A blanket of dark moved in from the horizon, blotting out the stars as it came. Drummond took her hand and they ran back towards the jeep.

He pressed the starter and moved away immediately, and in the same moment, great heavy drops started to splash against the windscreen. He pressed his foot flat against the boards, but it was no good. There was a tremendous clap of thunder overhead and the skies opened.

There was no time to put up the canvas tilt and he crouched behind the wheel, eyes narrowed against the stinging, ice-cold rain and Janet huddled at his side.

He drove into the courtyard at the mission, braked to a halt and they scrambled out and ran up the steps to the porch.

The thin silken dress was plastered to her body like a second skin and she shivered uncontrollably, laughing at the same time.

'That was marvellous, simply marvellous.'

'Better get out of those wet things,' he said. 'You'll catch your death.'

'You could use a towel yourself.' She took his hand.

'We'll go round this way. Father Kerrigan's probably gone to bed.'

They followed the verandah to the garden at the rear where the window of her bedroom stood ajar. She went in, turned up the lamp and found a spare towel.

'Do what you can with that while I get changed.'

'Like me to dry your back?' he said.

She gave him a quick push towards the window. 'Go on, get out of here.'

She pulled the curtain, peeled off her wet clothes and towelled herself briskly, still shivering. After a while, the shaking stopped and a warm glow spread through her body. She pulled on her dressing gown, tying the cord at her waist and went back outside.

Drummond wiped the rain from his head and face and hung the towel across the rail. It was bitterly cold by now and he stood there breathing deeply, taking the freshness into his lungs, filled with a strange inward restlessness.

'Feeling better?' she said quietly.

He turned slowly. Janet Tate was standing a few feet away by the rail and as lightning exploded, her face seemed to jump out of the night, the hair like a dark curtain to her shoulders. And she was beautiful, that was the thing which came to him with a sense of real wonder. Not just attractive, but beautiful, and he took two stumbling steps towards her, pulling her close.

The drumming of the rain on the corrugated iron roof increased into a solid roaring that seemed to fill her ears. She was aware of his strength, the arms crushing her to him and as her loose dressing gown parted, his lips found her bare shoulders, her breasts.

She leaned against him, caught in a strong current there was no denying, and was aware of his hands, fumbling at the cord of her dressing gown.

As it opened, she pulled away, struggling frantically. 'No, Jack, no!' He paused, head slightly forward, trying to see her more clearly in the half-darkness, and she pushed him away violently with both hands. 'Not this way, Jack! I'm not one of your kept women!'

For a long moment he stood there, almost invisible in the shadows, staring at her, and then, without a word, he walked rapidly away.

As another brilliant flash of lightning illuminated the empty verandah, Janet turned with a dry sob, went back inside and threw herself on the bed, anger and frustration sweeping through her.

Drummond had left the window of his bedroom open deliberately in spite of the cold. He lay in bed, propped against a pillow smoking a cigarette and thinking about Janet Tate as the rain drummed endlessly on the roof.

If that was the way she wanted it, then to hell with her. As he reached to stub out his cigarette in the ashtray on the locker at his bedside, there was a movement by the window, something stirred and Famia emerged from the shadows.

Her hair was unbound, hanging to her waist and she wore a loose silken robe fastened with a scarlet sash. There was a slight rustling of silk and as she moved into the narrow circle of the lamplight, the robe slipped to the ground.

She stood there for a moment, magnificent in her nakedness, breasts pointed with desire, hands flat against her thighs.

She moved forward quickly and his arms went out to enfold her, crushing her softness against him. He held her close, staring blindly out of the open window at the night as she moaned softly, digging her nails into his shoulder.

And after all, why not? This was one kind of answer and as good as any other.

In the darkness of the terrace, the old woman listened for a moment, then nodded to herself in satisfaction and crept quietly away.

It was close to dawn when he awakened, the sweat cold on his flesh. It was still raining hard outside and he hitched the blanket over his shoulders and turned into her warmth to sleep again.

Outside, the sound that had awakened him came nearer, the roar of an engine thundering through the rain. There was a squeal of brakes, boots running across the courtyard. Drummond got out of bed, reached for his dressing gown and padded to the window. As he moved out on to the verandah, Tony Brackenhurst stumbled on the top step and dropped to one knee, his face wild and strained in the light of the porch lamp.

'For God's sake, man, what is it?' Drummond demanded.

'Chinese troops,' Brackenhurst gasped. 'At Howeel. They over-ran my camp, slaughtered my men.'

'Chinese?' Drummond said. 'A patrol, you mean?'

'Hundreds of the bastards! Hundreds!' Brackenhurst sobbed.

Drummond stood stock-still for a moment and then pulled Brackenhurst to his feet. 'Have you told anyone else about this?'

Brackenhurst shook his head. 'No, I haven't had time.'

'Good, if word gets out too soon we might have a general panic and that plane of mine can take no more then fifteen in this kind of flying country.'

'That's what I thought,' Brackenhurst said.

'I bet you did. Now this is what we do. From here, we go to the mission to warn Father Kerrigan and Janet. We'll leave them my jeep and they can follow us in with Kerim as soon as they're ready.'

'What do we do then?'

'Come back to town in your Land Rover and break the news to the Khan. This might prove to him just how useless it was to rely on the border tribes for information.'

He returned to his bedroom and dressed quickly, pulling on fur-lined boots and his old naval flying jacket. Famia sat up in bed, the blankets clutched to her breast and watched him.

'When will you be back?' she said.

He took the Smith & Wesson .38 from a drawer, checked that it was loaded and slipped a box of spare cartridges into his pocket. 'God knows, but you'll be all right. You don't need me. You never did.'

He went out through the window and a moment later she heard the two engines break into life, one after the other, and the sound of them faded into the rain.

The door creaked open and the old woman crept in. 'Did you hear?' the girl said softly in Urdu.

The woman nodded and pulled the blankets aside. 'Come, girl, there is not much time and you know what must be done.'

Famia dressed quickly in an old pair of Drummond's drill pants and a white naval sweater that dropped over her slim hips. She pulled on slippers, nodded to her mother and moved out on to the verandah. A moment later, she was running through the quiet streets, head down against the rain.

Within five minutes, she came to a bungalow almost identical with Drummond's, ran up the steps to the verandah and knocked on the door furiously.

'Mr Cheung! Mr Cheung!' she called.

6

Action by Night

It was the rain which saved Brackenhurst, the sudden torrential downpour which turned a normally quiet mountain stream into a brawling torrent, in one place filling a dip in the road with a ford of ice-cold water.

He had spent a long, hard day in the mountains on his own, prospecting for ore specimens and now, on his way back to his base camp at Howeel, the sudden rush of water gleaming white and brown in his headlights caused him to stamp hard on the brake.

He got out, found a branch at the side of the road and poked it carefully into the water. It was at least four feet deep. He might be able to drive through, but on the other hand, if the damned thing bogged down, he'd had it. He climbed back into the Land Rover and reversed to the top of the hill, switched off his headlights and returned on foot.

The water was cold, damned cold, and it swirled around his thighs, numbing him to the bone. He floundered forward with a curse and found dry land again. Thank God the camp was no more than half a mile away.

He trudged along the dirt road, head down against the driving rain, the light from his electric torch reaching into the darkness. Somewhere up ahead he seemed to hear a cry and then another, confused shouting and the dull, flat report of a gunshot muffled by the rain. A second later came the deadly staccato of a machine gun.

He stood at the top of a small rise, a slight frown on his face as he looked down through the pine trees at the flickering light of the campfire. There was a flurry of movement, the noise of vehicles, a shouted command.

He moved off the road and went down through the trees cautiously until he was no more than twenty or thirty yards away from the camp, but above it on the hillside.

The hollow was alive with Chinese troops, little stocky peasants in quilted uniforms and peaked caps, shining Burp guns in their hands, and the heart seemed to freeze inside him.

He could see two of his men, Galur and old Abdul, standing beside the fire, hands raised high in the air. There was the sudden roll of an automatic weapon and Abdul fell back across the fire. Galur turned, burst through the ring of men and ran for the trees, head down. For a moment it seemed that he might make it and then a burst from a sub-machine gun drove him on to his knees.

The soldiers were calling excitedly to each other as they started to search the tents. More and more of them pressed into the camp and with a sudden roar, a troop carrier came down the road, followed by another and yet another, half-tracks at the rear for mountain warfare, instead of wheels.

Brackenhurst had seen enough. He turned and scrambled back up the hill. From somewhere to his left, there was a cry and a bullet passed through the trees severing a branch.

He put down his head and ran faster, one arm raised before his face to ward off flailing branches. A moment later he floundered across the ford and staggered up the hill to the Land Rover.

The engine was still warm and it burst into life with a surge of power when he pressed the starter. He reversed quickly, the tyres skidded for a moment, searching for a grip on the soft, crumbling edge of the track, and then they found it and he drove away rapidly.

Sitting at the wheel of the Land Rover in the courtyard of the mission and remembering what had happened at Howeel, Brackenhurst shivered involuntarily. He could hear the rise and fall of voices and looked out again at Father Kerrigan, at Drummond standing in the doorway, the old priest holding a lamp in one hand.

After a moment, Father Kerrigan went back inside, closing the door and Drummond ran down the steps and scrambled into the passenger seat.

'Right, let's get moving.'

'What did the old man have to say?' Brackenhurst asked as he drove away.

'What could he say? He's going to pack up as fast as he can and follow on in the jeep with Janet and the boy. No sense in staying to face what's coming. You know what they do to people like him.'

'What do you think the Khan will do?'

'What in the hell can he do except get out? He hasn't got a defence pact with India, which means they're going to sit tight on their side of the border, and if I know them, the Chinese will be smart enough to go just that far and no further.'

'But why?' Brackenhurst demanded. 'What in the hell can they possibly want with a dump like this? There's nothing here that's worth having.'

'You could say the same about the Aksai Chin and the Ladakh, but they moved in there and for the same reason. Prestige, a paper victory. The glorious Army of the People's

Republic takes back what was part of the Chinese Empire a thousand years ago. The fact that Balpur is a few thousand square miles of the most sterile territory on God's earth doesn't matter. It'll take the people's mind off the bad harvest back home.'

As they drove through the deserted streets, the sky was beginning to lighten over the mountains, and beyond the scattered, flat-roofed houses, grey and sombre, the river roared through the valley, swollen by the rain.

Later, at the palace, waiting for the Khan in the room where they had dined in what now seemed another age, another time, Drummond opened the french windows and stood on the terrace in the rain, listening.

The Khan was taking his time, but when he came in, he was wearing a khaki drill uniform, the medals above the left hand pocket, a splash of vivid colour in the grey morning. The major domo to whom Drummond had given the original message followed him with a decanter of brandy and glasses on a silver tray.

The Khan had dropped twenty years and

there was a new vitality in his step. 'It seemed to me that a drink might be in order, gentlemen. If what Ahmed has told me is anywhere near the truth, it may well be some considerable time before we have another.' The major domo filled three glasses, passed them round and left the room. The Khan toasted them silently. 'Now, Mr Brackenhurst, perhaps you would be good enough to tell me in your own words exactly what happened at Howeel.'

When Brackenhurst had finished, the old man turned to Drummond. 'What do you think?'

'I don't understand it,' Drummond said. 'Last time I was up there, there was nothing. Not a damned thing.'

'But that was ten days ago now, am I right?'

'What are you going to do?'

'I'm not sure. First I must confer with Colonel Sher Dil and Major Hamid. I have sent messengers already telling them both to meet me urgently at Army Headquarters.'

'Seventy-five men,' Drummond said. 'They won't go far and you can't rely on the tribesmen. They'll simply take to the hills and stay out of trouble. And I don't think the Indian Army will interfere.'

'A pessimistic view, but a correct one, I fear. How many can your plane take?'

'Not more than fifteen in this kind of country. I've got to get over those mountains, remember, and if we're going we must go quickly. Once the people get hold of the news we'll have a howling mob running for the airstrip. Father Kerrigan's the only other person I've told so far. We called at the mission and left my jeep. He's going to pack up as quickly as he can and follow us in with Kerim and Miss Tate.'

The Khan nodded. 'Good, my son must certainly be saved at all costs.'

'Then by my reckoning that gives us a possible passenger list consisting of yourself, Kerim, Father Kerrigan, Miss Tate, Brackenhurst here and Major Hamid, Colonel Sher Dil, too, of course, if he wants to come.'

'What about Cheung?' Brackenhurst put in.

'I was forgetting him.' Drummond turned to the Khan. 'Your Highness is probably well aware of Mr Cheung's true politics. God alone knows what the Reds would do if they got their hands on him.'

The major domo returned and handed the Khan a polished leather belt and holster containing a heavy British Army service revolver. He belted it around his waist, and smiled grimly.

'Then I think it is time to move, gentlemen. You may drop me at Colonel Sher Dil's headquarters. I suggest you then continue on to the airstrip and prepare the plane for immediate take-off.'

Outside, it was even lighter now, the sky a heavy uniform grey, the rain turning the dirt road into a quagmire as they drove down through the streets to the main square, braking to a halt outside the grim, barrack-like building that was Sher Dil's headquarters.

There was already a bustle of activity, and as the Khan got out the colonel came down the steps to meet him, Major Hamid at his shoulder. The Pathan glanced enquiringly at Drummond who held his thumb down and Brackenhurst drove away quickly.

Beyond the city, one or two tents were pitched, a flock of heavy, mountain sheep crowding in close to where a herdsman's fire already trailed grey smoke into the morning.

They bounced over the rutted track, skidding slightly in the mud, and went over the escarpment and down towards the airstrip.

The corrugated iron hangar looked ugly and forsaken in the grey morning and Brackenhurst braked to a halt a few yards away and nodded towards the airstrip itself, already a sea of mud.

'Not much of a surface to take off on.'

'Anything will do for a Beaver,' Drummond said. 'That's why they're so good for this kind of country.'

He took out his key, unlocked the padlock and pulled the doors wide, revealing the red and gold plane, and a quiet, precise voice said, 'Excellent, my friend, now move away, please.'

Cheung came round the corner of the building, an automatic pistol in one hand. In the other, he held a grenade. 'Going somewhere, Jack?'

'That was the general idea.' Drummond slipped his hands casually into the slanting pockets of his flying jacket, fingers closing around the butt of the Smith & Wesson. 'What is this?'

Famia moved from behind one of the doors and stood at Cheung's shoulder, looking faintly ridiculous in Drummond's sweater, which was by now so sodden and heavy with rain that it almost reached her knees.

'Well, I'll be damned,' he said.

Cheung smiled gently. 'No one is going anywhere, Jack. It was not in the plan.'

In one quick movement, he pulled the ring from the grenade with his teeth and tossed it inside the hangar. In the same moment, Drummond pulled the Smith & Wesson from his pocket and loosed off a wild shot that splintered the door behind Cheung's head, sending him running for cover.

Drummond turned and ran. Brackenhurst was already scrambling behind the wheel of the Land Rover. As the engine roared into life, the grenade exploded, hot air reached out to enfold Drummond, and the entire hangar seemed to sag.

As he jumped in, the Land Rover shot away, wheels churning the mud to liquid. Cheung moved into the open, firing steadily, the gun held in both hands with all the expertise of the marksman.

Brackenhurst took the Land Rover into a gully that slanted up the hillside giving them some kind of cover, and a moment later the Beaver's fuel tank blew up.

'And I hope that's taken the bastard with it!' he shouted.

They roared out of the gully on to a plateau which jutted like a shelf from the side of the mountain and gave a view of the plain below.

Cheung was standing a little distance away from the burning hangar looking up towards them and the girl lay face down in the mud a yard or two away from him.

Drummond was conscious of nothing, no anger, no pain. There was no time to wonder about what had happened or why. Survival was the thing from now on. The only thing that mattered.

Brackenhurst braked on the edge of the plateau seeking the safest way down, and beneath them the town was spread out like a map. Already people were stirring, moving in the streets in spite of the heavy rain.

'Bad news travels fast,' Drummond said.

In the square outside headquarters there was considerable activity. Three trucks moved up and parked outside and the drivers got down and stood in a small knot, obviously discussing what was happening.

Somewhere, Drummond was conscious of the noise, dulled by the rain, and then Brackenhurst screamed and pointed up into the sky to where a couple of planes flew out of the grey morning side by side, turned and broke formation, spiralling down like leaves falling from the branches of a tree.

The leading jet roared down the valley beside the river, banking so close to the mountain that for one frozen moment Drummond was able to distinguish the red stars on the wings.

'God in heaven, Chinese Migs!' Brackenhurst cried.

In the town below there were cries of alarm, people were standing in groups looking up at the sky, and as they scattered to run, the leading Mig swooped and fired its rockets, ploughing a double furrow across the square and scoring a direct hit on the first truck in line outside Sher Dil's headquarters. The truck's petrol tank exploded and debris and flames cascaded outwards to enfold the panic-stricken people who ran past.

The second Mig came in fast, rockets ploughing into the other two trucks and the flimsy mud and wattle houses beyond. As it swooped up into the grey morning, the leader was already banking, turning in to make his second run. He roared down, rockets hammering into the closely packed houses, and scored a direct hit on the ammunition store on the other side of Sher Dil's headquarters. A tremendous explosion sent a column of flame shooting up through the dark pall of smoke that was already enveloping the town as the second Mig followed the other in fast.

'Let's get moving.' Drummond slapped Brackenhurst on the shoulder.

Brackenhurst turned, his face very white, eyes staring. 'Down there? You must be mad.'

Drummond didn't argue. He dragged Brackenhurst across the seat and scrambled into his place behind the wheel. He took the Land Rover down the steep hillside and across the plain, and the smoke enveloped them so that he had to drive blind for several moments, swerving as a half-ruined house loomed out of the gloom. They bounced across a tangled mass of timber and masonry and turned into the main square.

A man ran out of the swirling darkness, his petrol-soaked body flaming like a torch. He vanished in the direction of the river. Someone screamed monotonously above the crackling of the flames and ammunition started to explode.

The Land Rover crunched across a burned and blackened body and Drummond braked hard. The screaming had stopped and the silence was somehow intensified by the crackling of the flames. On this side of the square there was hardly a house standing, and one end of Sher Dil's headquarters was a heap of rubble.

As Drummond jumped to the ground, Hamid staggered out of the entrance and leaned against the wall at the top of the steps, gasping for air, his uniform smouldering in several places.

Drummond ran up the steps and caught him as he started to fall. 'Easy does it. I've got you. What about the Khan?'

There was blood on Hamid's right cheek and he wiped it away mechanically. 'I don't know. Inside somewhere. The place is a bloody shambles.'

As Brackenhurst came up the steps to join them, the Migs came down the valley again. They grabbed Hamid between them and ran. As they staggered in through the door and hit the floor, cannon fire ripped up the surface of the square again, fragments of stone rattling against the shattered windows.

Drummond lay against the wall and waited while the earth trembled. Two soldiers sprawled on their faces in the centre of the room and Brackenhurst crouched in the far corner, eyes wide and staring.

The firing ceased as quickly as it had begun and the Migs faded into the distance, leaving only the smoke and the flames and the ruins behind.

Drummond got up and helped Hamid to his feet and Brackenhurst joined them. When he spoke, his voice shook a little. 'We've got to get out of here, Drummond. We've got to get moving.'

Drummond ignored him and turned to Hamid who was leaning against the wall, shaking his head from side to side like a wounded bull. 'Where was the Khan when the attack began?'

Something clicked in Hamid's eyes and he took a deep breath. 'In the radio room. Through here.'

The door was off its hinges and the room beyond was a shambles. Four or five men, dead or badly wounded, lay sprawled amid the wreckage and Colonel Sher Dil knelt by the window, the Khan in his arms. In one corner, the wireless operator still crouched by his seat, earphones in place.

Sher Dil was covered in dust, his uniform

singed and torn, but he seemed otherwise unharmed. Drummond dropped on one knee beside him and looked down at the Khan. The front of the old warrior's uniform was soaked in blood and when he opened his eyes, death stared out.

He gazed uncomprehendingly at Drummond for a moment and then his eyes seemed to clear. He reached out one bloodstained hand and held on tight, his mouth opening and closing as he tried to speak.

'Kerim,' he croaked. 'You will save Kerim? Your word on it.'

'It's all right,' Drummond said. 'I give you my word. We'll get the boy to safety, I promise you.'

The hand tightened on the front of his flying jacket, there was a hollow rattling in the Khan's throat and blood erupted in a sudden flow between his lips.

Drummond forced the bloodstained fingers apart and Sher Dil laid the Khan gently down on his back. The colonel removed his tattered tunic and covered the face and Drummond stood up and turned to Hamid.

'Any sign of the party from the mission?'

Hamid shook his head as Brackenhurst stepped in through the shattered doorway. 'The Land Rover's still intact, thank God. At least we've got transport. I hope that bastard Cheung rots in hell.'

Hamid turned to Drummond. 'What's he talking about?'

'It seems our good friend Cheung was working for the opposition all along. He got to the plane with a grenade before the Migs came in.'

'But we checked him out with Formosa,' Hamid said. 'He *was* a Nationalist agent, there can be no doubt about that. They communicated with him regularly. We looked the other way for obvious reasons, but we knew all about it.'

'Probably a double agent,' Drummond said and turned to Sher Dil. 'If you heard that, you'll know we're on our own and it's one hell of a step to the border. Have you managed to contact the Indian Army yet?'

'No, but the operator's still trying.'

There was the sound of sporadic gunfire

and they all turned and looked out of the window. A current of warm air had momentarily snatched away the veil of smoke revealing a small sugar loaf hill on the other side of the town. People were running towards the river, refugees from the town, men, women and children, a few herdsmen on horseback, their panic-stricken flocks rushing this way and that, getting in everyone's way.

A second later, the top of the hill was alive with troops in drab quilted uniforms. They started to fire as they swept on and the screams of the mob rose into the air like the smoke as they started to fall.

The tidal wave surged on, the soldiers calling to each other like hounds in full cry, running down the hill towards Sadar and the pall of smoke dropped back into place.

Hamid turned to Sher Dil. 'We've got perhaps five minutes before they get here. You must contact Indian Army Headquarters.'

A section of the roof crashed through into the room, scattering flame and sparks, and as Hamid and Drummond ran forward to stamp it out, Brackenhurst rushed outside. A second later, the engine of the Land Rover roared into life. When Drummond reached the door, it was already disappearing into the smoke.

Hamid cursed savagely and went back inside and Drummond stayed there, listening to the sound of the Land Rover fade into the distance, aware of the wireless operator's excited voice as he finally contacted Indian Army Headquarters.

The smoke swirled around him, touched with crimson, and the sickly-sweet stench of burning flesh was everywhere. In the great heat, things seemed to lack definition and nothing was real any more.

A bullet splintered the wooden framework of the door and several Chinese ran out of the smoke. He ducked inside as Hamid appeared at a shattered window and emptied a Sten gun, driving them back into the smoke.

Sher Dil turned from the radio and dropped the hand mike he had been using. 'From the sound of things we'd better get moving. Every man for himself, and try to get across the river. There's a village called Bandong ten miles due south on the road. We'll meet there.'

The rear door led into a fenced yard. It was strangely quiet and the smoke hung low in the heavy rain, reducing visibility considerably.

The wireless operator climbed up on the fence and swung a leg over. There was a sudden cry and a group of Chinese appeared about forty yards to the left. Several of them fired at once and he screamed and fell backwards into the yard, clutching his face.

Sher Dil scrambled through a gap in the fence and started up the slope and Drummond went after him, weaving desperately from side to side as the Chinese continued to fire. He was aware of Hamid hard on his heels, of Sher Dil disappearing over the rim of the escarpment.

He could taste blood in his mouth as he clawed his way up, slipping on the wet earth, and then the jagged rocks on the skyline loomed above him. He went over the top, head down, sobbing for breath and tripped over an outstretched foot.

He had one brief impression of Sher Dil sliding down the steep slope of shale to the river below, picking himself up at the bottom and plunging into the water, and then they moved out of the swirling smoke to surround him, small and misshapen in their quilted uniforms, each carrying a rifle that seemed too large for him, an old-fashioned sword bayonet on the end.

Hamid was lying on the ground a few yards away and a soldier stood over him, a foot on his neck. Drummond backed against a boulder and the brown peasant faces moved in on him.

7

Edge of the Sword

The town gaol was one of the few major buildings left undamaged by the attack, and from the small cell on the corner of the second floor, Drummond had an interesting view of the city through the barred window.

It was 10 a.m., four hours since the initial attack, but smoke still drifted across the stricken city through the heavy rain and a heavy grey mist moved up from the river and crouched at the end of the streets.

It was unbelievably cold and rain drifted in a fine spray through the bars as Drummond dropped to the ground. 'It's going to be an early winter this year. I feel it in my bones.'

'For us, a matter of academic interest only,' Hamid said from his bunk.

'You think so?'

There was the sullen chatter of a machine gun from down by the river and Hamid smiled bleakly. 'There's your answer. Nothing like cutting down on the opposition. They haven't stopped since this morning.'

'Then why have they let us last this long? Why the special treatment?'

There was no time for a reply. A key grated in the lock, the door opened and a small sergeant stepped in, flanked by two privates armed with sub-machine guns. Hamid got to his feet and the sergeant shook his head.

'Not you, this one.'

They pushed Drummond into the corridor before he had a chance to say anything, and the door clanged into place with a grim finality.

The sergeant turned without a word and started along the corridor and Drummond followed, the two privates bringing up the rear. They mounted a flight of stone steps to the top floor and halted outside a door. The sergeant knocked, listened for a moment and then led the way in.

The room had once been the governor's office. The walls were hung with Bohara rugs, sheepskins covered the floor and logs burned in the large stone fireplace.

A Chinese officer stared down into the fire, one foot on the hearth, and tapped his booted leg with a leather swagger stick impatiently. The heavy greatcoat with the fur collar which swung from his shoulders carried the epaulets of a full colonel.

He turned and examined Drummond calmly. 'You don't look too good, Jack.'

'No thanks to you, you bastard.'

'Nothing personal, Jack. We just happen to be on different sides. Regrettable, but true.'

'What are you, Military Intelligence?'

'That's right.'

'Am I allowed to ask for how long you've been making a monkey out of them back on Formosa?'

'I've never been to Formosa,' Cheung said. 'The Nationalists *did* have an agent called Cheung and they *did* send him on a mission to Nikkim. He got as far as Singapore. I took his place from there.'

'What about the guns I flew into Tibet and Moro and his boys? All that was a fake, too, I suppose?'

'An elaborate pretence which enabled me to communicate constantly with my superiors to help pave the way for the regaining of what has always been legally a part of the Chinese Empire by the Army of the People's Republic.'

'I can do without the speeches,' Drummond said. 'Where did Famia fit in?'

'She and her mother were of no particular importance. I paid them well to keep me informed of your movements.'

'You speak in the past tense.'

'Only where Famia is concerned. She was struck in the head by a piece of sharpnel back at the airstrip.'

Remembering the months of pretence, Drummond dismissed her with no particular regret. 'Do you really think you're going to get away with this?'

'Why not?' Cheung said. 'India will not interfere. She is interested only in maintaining the status quo and the two countries have never had a mutual defence pact. In the United Nations, there will be an emergency session, they will talk far into the night and do precisely nothing. No one wants to rock the boat, Jack. A phrase you taught me.'

'You've got it all nicely worked out, haven't you?'

'Except for one rather important detail. The Khan is dead, which is something of a convenience, but these are a superstitious people, and to them the Khan is priest as well as king. Prince Kerim can be his only successor.'

'And as such, an obstacle to the setting up of a People's Republic.'

'Not at all.' Cheung smiled briefly. 'With our guidance, he could be of great help to his people. He could lead them along the true path.'

'Now I've heard everything,' Drummond said.

'Good, then perhaps you will be sensible enough to help me in this matter. After all, you always did have an eye to the main chance. Where is the boy?'

Drummond stared at him in astonishment. 'You mean you don't know?'

'He is not at the mission. Neither are Father Kerrigan nor the American girl. My men have spent hours checking the crowds and searching the immediate area of the city.'

'And you expect me to help?'

'I know you went straight to the mission from your bungalow after Brackenhurst brought you news of the invasion. Famia told me.'

Drummond decided to take refuge in the truth. 'That's right. We left my jeep and told Father Kerrigan to pack up as quickly as possible and meet us at the airstrip. Your men moved too fast for him, I suppose.'

'But Brackenhurst also is proving difficult to locate. You arranged an alternative plan in case of trouble. I know you, Jack. I know how your mind works.'

'Can I go now?'

'You might find it wiser to co-operate. I could make things easier for you.'

'Do me a favour, for God's sake,' Drummond said. 'That's like a line from a bad play.'

Colonel Cheung stared at him, apparently calm, the leather swagger stick beating against his right boot. 'All right, take him away, Sergeant,' he said abruptly in Chinese. 'Bring the other one.'

Drummond paused in the doorway and shook his head. 'Now you really are wasting your time,' he said.

Chinese headquarters had been set up at the palace and the commanding officer, General Ho Tsen, stood on the terrace and looked out over the garden. He seemed far from happy and paced up and down impatiently.

There was a slight cough from behind and he turned to find Cheung standing in the window. 'You have found him?' he demanded eagerly.

'I'm afraid not, General.'

Ho Tsen slammed a hand hard down on the balustrade. 'This is your direct responsibility, Colonel. I expected to find the boy ready and waiting when I arrived.'

'It seems that the priest and the American

girl left the mission with the boy shortly before our men arrived. We've confirmed this by questioning herdsmen at a camp up-river. I've just had a report that their jeep has been found abandoned ten miles north of here at a village called Quala. There was a vehicle ferry there which has apparently been destroyed so I'm assuming they've crossed by boat. A Land Rover belonging to the man Brackenhurst has been found in the same place.'

'Has the patrol gone after them?'

'Unfortunately there were no other boats. The village was quite deserted. Obviously the entire population had crossed over. Since then the level of the water has risen with the rains.'

'Is there any place where the river may be crossed with vehicles?'

'Certainly not here. The current was always too swift for a ferry.'

Cheung spread out a map on a wrought iron table. 'Twenty miles north of here at Kama. The river is very wide and shallow there. We could cross in half-tracks.' His finger traced a line to the border. 'There is only one road to India and they have no transport, remember. We should catch them easily. They must stay on the road. The priest is an old man and in any case, with the woman and the boy, he couldn't hope to get through the mountains on foot by any other route.'

Ho Tsen nodded. 'I hope so, for your sake. Pekin will not be pleased if you fail. I will also send other patrols south on this side of the river in troop carriers. They should find boats sooner or later. Once across, they can proceed on foot and cut the road ahead of you.'

'An excellent idea.'

Ho Tsen put a cigarette in his mouth and leaned to the match Cheung offered. 'One thing worries me. What if the priest had an alternative plan? Perhaps a vehicle waiting on the other side. It would explain why they did not drive north to Kama and attempt to cross in the jeep. This man Drummond you spoke of? You are certain he knows nothing?'

'He is a difficult man to be sure of and

the Pathan is as stubborn as the breed usually are.'

'You have exhausted the accepted methods?'

'They take time, General, and in any case, Drummond must be preserved for a more searching examination in Pekin.'

'Why is this?'

'He is known to have worked for British Intelligence.'

'I see!'

Cheung hesitated. 'I would like to have one last try before leaving in case they do have information of value. A small subterfuge which often has remarkable results.'

'Which sounds interesting,' General Ho Tsen said. 'I think I shall accompany you, Colonel. Let us hope I'm not wasting my time.'

The wind across the river was like a bayonet in the back and Drummond shivered as it cut into him. He flexed his hands to ease his cramped muscles and winced with pain as the wire that was twisted about his wrists bit into his flesh.

Hamid was next to him and on the other side one of Sher Dil's soldiers in tattered uniform was silently weeping. Every few moments the man coughed and a trickle of blood came from his mouth. After a while, he slumped on his face and lay motionless. The guards standing talking a few yards away took no notice.

Two troop carriers, half-tracks biting into the mud, drove up and parked thirty or forty yards away, each containing a dozen men and a heavy machine gun mounted on a pivot.

Drummond eased back on to his heels and looked along the line of kneeling men. There were at least thirty of them, mostly Sher Dil's soldiers with a few tribesmen who'd been caught carrying weapons. In his mind's eye he saw them keel over one by one as the machine gun curved in an arc and finally reached him and he shuddered.

A jeep drove up and parked behind the troop carriers and Hamid said quickly, 'We've got company, Jack.'

Cheung walked across the broken ground

towards them, General Ho Tsen at his side. They paused a few yards away and the General said calmly, 'These are the men?'

Cheung nodded. 'They both speak Chinese.'

'Excellent.' The General came closer. 'Let us not waste any more time, gentlemen, I find this rain most unpleasant. We wish to know the whereabouts of the Catholic priest and the young Khan. If you are sensible and help us, I will see that you are well treated. If not . . .'

Drummond and Hamid stared up at him without speaking and Cheung sighed with exasperation. 'You're a damned fool, Jack,' he said in English. 'You always were. We've found the jeep at Quala which means they've crossed the river. They won't get far, I promise you.'

He and the General turned and walked back to the troop carriers. Ho Tsen climbed into the shelter of the jeep and Cheung looked up at the sergeant who stood beside the heavy machine gun in the first troop carrier.

'You have your orders. Stop firing when you reach the Indian and the Englishman.

If you harm either of them, I'll have your head.'

He climbed into the jeep beside the General and Ho Tsen smiled and offered him a cigarette. 'You were quite right, Colonel. This should prove most interesting.'

Drummond stared at the sodden earth, numb with cold and waited for what was to come. He wondered about Father Kerrigan and Janet and the boy, somewhere on the other side of the river in the mist, and prayed that the old man would have the sense to keep on the move. But the Indian border was a long way off, all of three hundred miles.

A burst of shrill, girlish laughter came from the Chinese and he stiffened. They strutted towards the line of prisoners, their thin voices bird-like on the wind and Drummond dropped his head and waited.

A boot thudded into his chest and he rolled on his face and fought for breath. The wire was torn from his wrists and a kick in the side drove him to his feet. The Chinese soldier grinned amiably and held out a spade.

Drummond glanced once at Hamid and

they started to dig. The soil was soft and sandy and lifted easily. Beside them the other prisoners worked silently, and as Drummond bent to his task, he knew with a feeling of utter hopelessness that it wouldn't take very long.

The rain increased into a heavy downpour and the Chinese turned and ran to the shelter of their vehicles leaving one man on guard, a sub-machine gun crooked in one arm.

The trench was now a couple of feet down and Drummond wondered how deep they wanted it. Six feet was the statutory requirement for a grave back home, but it was unlikely that the Chinese bothered about such niceties.

He leaned on his spade for a moment and Hamid moved closer. 'I don't suppose we've got much longer,' Drummond said.

Hamid glanced once over his shoulder at the mist rolling up from the river. 'Not if I can help it. Any good at the hundred yard dash, Jack?'

Drummond frowned in bewilderment. 'What in the hell are you talking about?'

'This,' Hamid said crisply and slapped him heavily in the face.

As Drummond staggered back, momentarily dazed, the guard hurried across to see what the disturbance was about. He leaned over the trench, the sub-machine gun pointed threateningly and Hamid swung the edge of the spade against his neck. The man fell into the trench without a sound.

The rain was now a heavy grey curtain that almost shrouded them from the troop carriers and the jeep. Hamid snatched up the guard's sub-machine gun, scrambled out of the trench and ran towards the river. Drummond went after him, slipping and stumbling in the mud.

Behind him he heard a cry and glanced over his shoulder. The other prisoners were strung out in a ragged line, running for dear life. Beyond them, the first Chinese had already reached the trench, firing as they advanced, and one of the troop carrier's heavy machine guns opened up above their heads.

The river was very close now and he increased his pace as he smelled water. A bullet plucked at his heel and he tripped and fell heavily. Hamid was beside him in an instant. He dragged him to his feet and together they stumbled down the slope to the water.

The river was in a sullen, angry mood. It ran smoothly through the heavy rain, but sudden swirls on the surface indicated dangerous currents and the speed with which tree branches drifted by argued against any attempt to reach the other side.

There was a rattle of stones and earth behind and one of Sher Dil's soldiers ran past, his face purple with effort so that a scar which stretched from one eye to the corner of his mouth gleamed whitely. He plunged into the water and began to swim furiously.

In a few moments, the swift current had carried him out of sight into the heavy rain. Others followed, some bleeding from wounds, crying with fear as they stumbled down the slope and flung themselves into the river.

'They won't last five minutes,' Hamid shouted. 'The water must be near freezing.'

A bullet landed at their feet, showering them with dirt. He turned as four Chinese appeared at the top of the slope and swung the sub-machine gun in a wide arc, firing from the hip. Two of them crumpled to the ground.

As their companions dropped behind an outcrop of rock, several more appeared on the skyline and Hamid drove them back with a long burst that emptied the gun.

As one of the troop carriers came into view, Hamid tossed the useless weapon to one side and they ran into the river and splashed through the shallows towards a line of dense, thorny bushes that grew down into the water. Bullets churned around them and then they were waist deep and hidden by the rain.

The water was bitterly cold and Drummond could feel it eating its way into his bones. They could hear the cries of the soldiers coming nearer and began to move further downstream aided by the strong current.

The land curved out for about fifty yards making a natural breakwater, imprisoning a floating mass of smashed trees and branches. They pushed towards it and the current, taking pity on them, swept them into the safety of the floating jungle.

They rested side by side, holding the branches of a tree, gasping for breath. Voices came from the shore and half a dozen soldiers appeared, pushing their way through the bushes along the water's edge.

They were no more than ten yards away from the shore and through the branches Drummond could see the Chinese clearly, the peaked caps pulled down over the eyes, the red star prominent, the shining Burp guns and rubber, knee-length boots.

The breathless minutes passed slowly and the cold gradually numbed their limbs. The soldiers appeared to be having a conference. After a while, they split into pairs and disappeared into the driving rain.

'Now what?' Drummond said.

'Only one thing for it,' Hamid said, his lips blue with the cold. 'We'll have to try to get across on one of these logs. They'll be swarming around like flies on this side of the river. We wouldn't last five minutes.'

He let go of the tree and splashed to the

next one, progressing slowly through the floating mass and Drummond followed. When they reached the edge, they found a large tree that was already swinging out into the river, straining to be free.

Hamid pulled himself into the branches and Drummond said, 'I'll try to guide it from the other end.'

He lifted a foot from the water and pushed against the next tree. There was a snapping of branches and the tree lifted into the current. In a few moments they were drifting rapidly away and the floating mass of trees and the promontory disappeared into the mist.

Drummond soon found that it was impossible to guide the tree. It went with the current and his feeble kicking had no effect. He gave up the struggle and tried to heave himself into a more secure position, but his frozen limbs refused to help him. He trailed helplessly through the water, arms hooking over a projecting branch and gradually all discomfort and pain left him.

* * *

When Cheung scrambled up the steep bank from the river's edge, he found General Ho Tsen still sitting in the front of the jeep, a cigarette in a long jade holder between his teeth.

'Well?' the General demanded.

Cheung seemed tired. 'As yet there is no sign of them, General.'

'A small subterfuge which often leads to remarkable results,' said Ho Tsen. 'Wasn't that what you promised me?'

Cheung wiped rain from his face mechanically. 'What can I say?'

'Nothing,' Ho Tsen told him. 'That would be much better. As it happens, in such weather it is more than likely that Drummond and his friend are already floating face down somewhere out there in the flood. In any case, I shall take charge here, Colonel. Take your men and go north to Kama. Cross the river and bring me back the young Khan.' He paused and neatly ejected the end of his cigarette from the holder. When he looked up again, his eyes were cold. 'Without him, there would be little point in returning at all. You follow me?'

Cheung stood there in the rain, staring at him for a moment, his face quite white. He seemed to pull himself together, saluted, turned and clambered up beside the driver of the first troop carrier. A moment later, the two vehicles moved up the slope, their tracks spurning the wet earth and disappeared into the mist.

8

Forced March

Vaguely through his numbed mind, Drummond became aware that something was digging into him. After a while, he realised that another large tree had drifted into them. They floated together, branches intertwined, their combined weight considerably slowing down the speed at which they were travelling.

Hamid was still secure in his perch amongst the branches and after a while he called, 'I can see the other side. The river must be narrowing.'

Drummond turned his head. Through the torrential rain, the opposite bank was just visible and it seemed to draw nearer every moment. The water became rougher and trees and flotsam of every description raced through the white-capped waves.

Suddenly, the bank was very close and seemed to increase in size as the river churned through a deeper and narrowing channel. There was a sickening, body-shaking jar and the tree grounded.

Drummond heard a cry and saw Hamid flung into the water. He unhooked his cramped limbs and found that he could stand waist deep. He forced his way along, pushed by the current, and caught Hamid by the belt before the river took him. There was a crashing sound behind and, as he turned, the trees were lifted by a sudden swell of the water and swept away again.

The water boiled around them as they braced themselves against the current. Slowly they forced their way to the steeply shelving banks and scrambled to temporary safety. They lay face down, their battered bodies heaving as they retched up river water.

After a while, they got to their feet and clambered up the mud bank away from the river. They stood looking across the river through the mist, listening. Hamid was shaking with cold, his uniform moulded to his body although strangely enough, his turban was still intact.

'Sooner or later they'll get men across by boat,' he said. 'They're bound to find an odd one or two missed by the refugees.'

'But they'll be on foot, just like us,' Drummond reminded him. 'The nearest place they have a hope of crossing with vehicles is Kama and that's twenty miles north from here. The shallows there could well be impassable because of the rain.'

'Well, one thing is certain,' Hamid said with a savage grin. 'There's only one road out to India and there's only one way we're going to get to it.'

They began to walk south through the rain, slowly, because the ground was fast turning into a quagmire. Drummond found it an effort to lift one foot in front of the other, and after a while found himself falling behind the hardy hillman.

They moved into a grey impenetrable mist that shrouded them completely from the outside world. Nothing existed now except the two of them and the rain and Drummond stumbled on through the mud, wondering what he was doing here and where it was all going to end.

It was perhaps half an hour later that he became aware that Hamid was calling to him. He was standing on top of a small hill about fifty yards away. When Drummond joined him, he saw a herdsman's hut in a small hollow below.

There was no sign of life and they moved cautiously down into the hollow. Drummond didn't feel tired any more. He didn't feel anything. He knew he was alive and that was about all.

It was a poor place of mud and wattle construction and a thin tracer of smoke lifted through a hole in the straw roof. Hamid opened the door and led the way in.

The fire on the stone hearth was banked with earth and smoke drifted in a heavy layer against the ceiling. It was filthy and it smelled and Drummond knew the place was very probably lousy with fleas as well, but it was warm and dry, and at the moment that was all that mattered.

He raked the soil from the fire and brought wood from a pile in one corner. Hamid rummaged amongst the sheepskins at the back and came up with a couple of stone jars.

He brought them to the fire with a grin. 'Goat's milk and cheese. Pretty rancid, but good for the constitution.'

'At the moment, I could face anything except going out there in these wet clothes again,' Drummond said.

He built the fire into a great, roaring pyramid and Hamid gave the sheepskins a shake. 'God alone knows what we'll get from this lot.'

But it didn't matter, nothing mattered except that it was warm and the fire was hot on the skin. Drummond crouched there watching the steam rise from his clothes suspended from the ridge pole, a sheepskin around his shoulders, and after a while he slept.

* * *

He awakened slowly and stared through the dim grey light at his clothes hanging from the ridge pole of the but, wondering where he was. After a while he remembered and sat up.

Hamid squatted on the other side of the fire. He was wearing his uniform again and grinned. 'How do you feel?'

'Bloody awful!' Drummond stretched his arms and blood started to flow through cramped limbs. 'How long have we been here?'

'A couple of hours, that's all. Must be about two o'clock. We'd better get moving. Your clothes are pretty dry by now. Better than they were, anyway.'

Drummond started to dress and Hamid peered outside. 'From the looks of it, this rain is never going to stop. I think it'll turn to snow before it does.'

'As if we haven't got enough to worry about.'

Hamid shrugged. 'The weather should help if anything. It makes things just as difficult for the Chinese.'

Drummond moved to the entrance, zipping up the front of his flying jacket and looked out. The rain was lancing into the earth with steady force and a slight mist rising from the cold ground combined with it to reduce visibility to a few yards.

'I think you're right about the snow.'

'Which means we've got to move fast. We can't be more than seven or eight miles from the road. Anyone else who got across the river is bound to move in the same direction. They've no other choice.'

'You're thinking of Father Kerrigan and Janet?'

'Or Sher Dil, but the Chinese will follow the same route once they get across and we must keep ahead of them. If we can only reach the village Sher Dil mentioned, Bandong, and get horses, we might stand a chance.'

He picked up a couple of sheepskins and tossed one to Drummond. 'Better wear that over your shoulders. It'll keep out some of the rain.'

In the same moment, he drew back from

the entrance, a finger to his mouth and dropped to one knee.

They crouched side by side, soundless and waiting. At first there was only the savage drumming of the rain and then Drummond heard it. A slipping, stumbling sound of feet trailing through the wet ground outside.

As the steps approached the hut and paused, Hamid launched himself through the entrance. There was a sudden splashing through the mud outside, the sound of a blow.

Drummond went after him, fists ready, but there was no need. Hamid stood over the huddled figure of a man who crouched in the mud. He grabbed a handful of hair and jerked the head back savagely. A great scar ran from the man's right eye to the corner of his mouth. The tattered remnants of a khaki uniform with corporal's stripes on the right sleeve still clung to his wiry body.

'It's the one who plunged into the river ahead of us,' Drummond said. 'You remember? He's one of Sher Dil's men.'

The man's face split into a wide, impudent grin. 'You know me, Major Hamid. Ahmed

Hussein, Corporal in Number One Section.' His English was almost perfect, but with a slight, sing-song accent. 'Drummond Sahib, I have seen many times.'

Hamid started to laugh. 'I know this one, all right. One of the greatest rogues you'll ever meet in your life. An old Indian Army man, Khyber Rifles, wasn't it?'

'That's right, sahib.' Ahmed got to his feet and indicated the row of medal ribbons above his left breast. 'D.C.M. from King George himself, sahib.'

'Probably bought in the bazaar at Peshawar,' Hamid said. 'But he's an Afridi. They're good fighting men.'

They went outside and Ahmed crouched over the fire, warming his hands. 'What are we going to do with him?' Drummond said. 'We can't afford to wait for him to dry off.'

'No need, sahib.' Ahmed picked up another sheepskin. 'This will do fine. The cold is nothing to me. Hardship is nothing.' He grinned hugely. 'I'm an Afridi.'

'Which also means liar, cheat and rogue,' Hamid said. 'There's goat's cheese in one of the jars. If you're hungry, you'll have to carry it with you and eat on the way.'

'Where do we go, sahib?'

'To the road, where else? The road out of this accursed country. Colonel Sher Dil told us to meet at Bandong if we managed to cross the river. Do you know it?'

'About eight miles south, sahib. I take you there.'

When they climbed out of the hollow, Drummond paused for a moment and looked down at the small hut, the smoke rising into the air. Somehow it represented security and safety and now he was moving into the unknown again. He shivered and hurried after his two companions.

For the first quarter of a mile, Ahmed trailed at the rear scooping handfuls of the soft cheese from the jar, devouring it avidly with groans of delight. Finally, he tossed the jar to one side and ran ahead to take the lead.

Drummond kept a pace or two behind Hamid. The world was a few cubic feet covered on all sides by walls and a ceiling of mist and rain and they were the only inhabitants.

They had been marching for about half an hour when he stumbled into Hamid who grabbed his arm. 'Listen for a moment.'

Ahmed joined them and they stood in a small group, strange figures in their sheepskins, streaming with rain and somehow symbolic like a piece of modern sculpture.

'I thought I heard firing,' Hamid said and at that moment it sounded again, a faint echo to the west.

'Sounded like a machine gun,' Drummond said.

Again, there was a faint, deadly echo of small arms fire and then there was silence.

'Probably back across the river,' Hamid said. 'We're still moving parallel with it, remember, but I think one of us should scout ahead from now on.'

'I will go, sahib,' Ahmed said with a grin and ran into the mist.

They commenced to march again. Drummond's senses were on the alert for danger at first, but gradually he succumbed to his surroundings. There was a safety, an anonymity about the rain and the mist that was vaguely comforting.

He withdrew into himself, an old trick, forgot about fatigue, discomfort, the danger of his present situation. He didn't even feel fear when Ahmed suddenly emerged from the mist and ran towards them.

Hamid grabbed hold of the Afridi and steadied him. 'What is it?'

'There is a village up ahead, sahib.'

'Good, lead the way.'

He walked into the mist and they followed him. Drummond found that he was sweating a little, the ground sloped and then dipped suddenly as they descended into a large hollow.

The houses loomed out of the mist. There were no more than six of them, poor, mean places of mud and wattle like the herdsman's hut scattered alongside the banks of a small stream.

They went forward quickly and Drummond was aware of the acrid smell of woodsmoke on the damp air. Ahmed opened the first door and went in. He reappeared a moment later.

'Empty, sahib, everything gone.'

He ran along the line of huts, opening the crude, wooden doors and finally came back to meet them despondently. 'Picked clean, sahib. Picked clean.'

Hamid looked in through the door of the nearest house at the embers of the fire which still glowed on the hearth. 'I said bad news travels fast, didn't I? They've gone, every last one of them. Horses, livestock, the lot. Taken to the hills I suppose, to wait things out and see what happens.'

Ahmed looked at them enquiringly. 'We move on now? Nothing for us here.'

'That's right, Ahmed,' Drummond said. 'Nothing for us here.'

They moved up out of the hollow and started to march again. The rain-soaked earth made the going very heavy and the ground itself was boulder-strewn and very difficult so that they had to pick their way with care.

Gradually a change became noticeable. The air seemed colder and drifted steadily into their faces and the ground began to slope steeply. They paused to take stock of the situation.

'We must be coming to the edge of the rift valley,' Hamid said. 'And that means the road can't be far away. We should cut across it in another mile or so.'

They started to make their way down the hillside. The ground began to fall away until at times, they were compelled to climb very carefully, feeling for handholds.

Finally, they found themselves on the lower slopes and the going was easier over rough, moss-covered ground. Ahmed moved ahead again and was soon lost to sight, for as they descended through the rain, the mist became thicker until visibility was almost nil.

It was Drummond who heard the motor. He stopped quickly and called to Hamid. They both stood there on the hillside listening and heard the sound of truck engines.

Ahmed came running out of the mist. 'Bandong just below in the valley, sahib,' he said to Hamid. 'Four trucks stopping there. Big ones, sahib, I think they are ours.'

'What do you mean, ours?' Drummond said.

'Army trucks, sahib. Convoy from India making its way to Sadar.'

'He's right,' Drummond said. 'I'd forgotten about that. Don't they make the run once a month?'

'Only one difficulty,' Hamid said. 'If it is the usual convoy to Sadar, then it's going in the wrong direction.'

'Not if they'd heard what's happened.'

They covered the rest of the distance quickly, running and sliding down the slippery slopes until they came to a boulder-filled stream bed. On the other side they scrambled up on to a dirt road, and Ahmed motioned them to silence as a flat-roofed house loomed out of the mist.

'Bandong,' he whispered.

The truck engines had stopped and the whole world seemed to have died with them. A vague unease stirred in Drummond and then he heard the voice, the rough, familiar Irish voice, and ran forward between the houses scattered on either side of the road.

Four trucks were drawn up in a line, old Bedford three-tonners, pointing south towards India. Father Kerrigan stood bareheaded in the rain talking to a tribesman in sheepskin coat and fur hat who held an old .303 Enfield rifle in one hand and the bridle of a rough hill pony in the other.

A stone rattled under Drummond's foot and they swung round. The hillman was Colonel Sher Dil.

'Well, praise be,' Father Kerrigan said softly.

The door of one of the trucks opened and Janet Tate dropped to the ground. She was wearing the same clothes she had worn on the flight in, fur-lined boots, cord pants and the sheepskin jacket Drummond had provided for her, but he didn't really notice these things. Only her eyes and the deep incredulous joy in them as she ran towards him.

9

Council of War

A corporal and three privates walked forward slowly, curiosity written on their faces, and behind them, lagging slightly, his left arm heavily bandaged, came Tony Brackenhurst.

'We didn't expect to see any of you again,' Father Kerrigan said. 'The Chinese arrived so quickly that we only got out of the mission by the skin of our teeth. I drove up-river to Quala and found that the headman had already had the vehicle ferry destroyed to prevent the Chinese crossing the river with transport. Everyone in the village was being ferried over by small boats in relays.'

'Mr Brackenhurst arrived while we were waiting our turn,' Janet continued. 'He was pretty badly burned. He told us what happened at Sadar. He thought he was the only one to get away.'

'So he was for a while,' Hamid replied calmly.

Brackenhurst looked very pale and swayed slightly, groping for the side of the truck to steady himself. Two of the soldiers moved to catch him and Father Kerrigan said, 'I think you ought to lie down again, my boy, you don't look too good. Will you see to him, Janet?'

Brackenhurst stumbled away between the two privates, Janet walking beside them, and the priest turned back to the others.

'I don't think I've ever had a greater surprise in my life than I did ten minutes ago when this tribesman here emerged from the mist and turned out to be Sher Dil.'

'I arrived on foot about four hours ago,' Sher Dil said. 'When I told the villagers what was happening, they decided to move into the mountains while they still could. They wanted me to go with them, but I'd told Drummond and Major Hamid to meet here if they managed to get across the river.' He grinned. 'I was beginning to think you weren't going to make it.'

'We very nearly didn't,' Hamid said. 'They insisted on our staying for a while. You'll be interested to know, by the way, that friend Cheung is an Intelligence Colonel.'

'God bless my soul,' Father Kerrigan said. 'Are you sure about this?'

'We've the best of reasons for knowing, Father,' Drummond said. 'How's Kerim?'

'Taking it all surprisingly well. Of course, he's not had things too bad as yet. After crossing the river we were in an ox cart for nine or ten miles, but then we met the convoy. As soon as Corporal Nadin heard our story, he turned round at once. He didn't have much choice. He couldn't have gone any further.'

'Does he know about his father's death?'

'So it's certain? Brackenhurst seemed to think so, but I kept hoping he might have been mistaken.' Father Kerrigan sighed. 'No, I've told the boy nothing. Later, perhaps, when we're safe across the border.'

'If we can get there, Father. A debatable point at the moment.'

Ahmed splashed towards them through the rain, two tin mugs in each hand. 'Tea, Colonel?'

'So, you survived, you rogue?' Sher Dil said in mock anger. 'Am I never to be rid of you?'

'As Allah wills, Colonel.'

Ahmed grinned impudently. He wore brand new leather combat boots and a quilted khaki parka of the kind specially issued for winter warfare, the fur-lined hood pulled up over his head.

'Where did you get the clothes?' Drummond asked.

'One of the trucks was carrying general equipment for the army, sahib. There is still some left although we unloaded most of it back there on the road to make room for the women and children.'

'Women and children?'

'Refugees we found on the road. We could not leave them for the Chinese.'

'Tell Corporal Nadin to bring me a map,' Sher Dil said.

They squatted on the verandah of the nearest house to drink their tea and Nadin, a thin, sinewy Indian with a brown face and long black moustache, brought the map.

Sher Dil unfolded it. 'Three hundred miles to the Indian border and one road out – this one. The usual way to cross the river with transport was by the ferry at Quala, but according to Father Kerrigan, the villagers have burned it.'

'It might be possible to cross in the shallows at Kama,' Drummond said. 'Especially with half-tracks, and they've got those.'

'Do you think they'll try?' Father Kerrigan said.

Hamid nodded. 'I'm afraid so. They want the young Khan, Cheung made that quite plain. A puppet to sit on the throne of Balpur, a mouthpiece for the People's Republic. As Sher Dil says, there's only one road out. They're certain to follow.'

'Then we must keep moving. We have a good lead.'

'Only for a time.' Sher Dil ran his finger along the course of the river. 'Here, seventy miles south of Sadar is a village called Huma. If the Chinese get their hands on the boats there, they can put men across.'

'But not vehicles.'

'That is true, but see how the river swings to follow the valley. They would be no more than ten to fifteen miles from the road, no distance for active, well-trained troops.'

'So you think they may try to cut the road ahead of us?' Drummond said.

Hamid shrugged. 'I don't know who their commanding officer is, but that's what I'd do if I was in his place.'

'Then the sooner we're on the move, the better.'

Sher Dil looked up at the leaden sky. 'About two hours of daylight left. We can go a long way in that time.'

'You don't think we should push on through the night?'

'On this road?' Hamid laughed harshly. 'It would be suicide in vehicles like these. Much better to camp at a suitable spot and move on at first light. The Chinese won't have anyone across the river yet, there hasn't been enough time. We'll have a good start on them.'

Sher Dil got to his feet and turned to Corporal Nadin. 'What about petrol?'

'There is plenty, Colonel, enough for all the trucks.'

'Why not dump two and push on in the others?' Drummond said. 'Plenty of room for all of us if we unload.'

Sher Dil laughed and made a sweeping gesture that took in all four of the Bedfords. 'Look at them. Twenty years old if they're a day. They've been running since the Burma campaign and it shows.' He turned to Nadin. 'How often do they break down?'

The corporal shrugged. 'All the time, Colonel. First one thing and then another.'

'That settles it. We push on with all four. If one breaks down, we still have three left and so on. One of the damned things is bound to last out to the border. In any case, the ammunition they carry may prove useful.'

The three privates had been standing in a little group a yard or two away listening to this conversation and as Corporal Nadin turned to move away, one of them grabbed his sleeve and muttered something quickly.

Sher Dil frowned and stepped down into the mud. 'What's going on?'

Nadin turned, indecision on his face. 'Two of the men, sir, Piroo and Yussuf. They are local men. Their wives are in Sadar. They would rather stay. They do not wish to return to India.'

Only the rush of the rain and the rattle of water in the stream bed on the other side of the village disturbed the silence after he had finished speaking.

Father Kerrigan looked worried and Hamid was quite composed, ready for anything. When Drummond glanced quickly at Sher Dil, the colonel's face had turned pale with anger and the eyes blazed fire.

'For a soldier who disobeys an order in the face of the enemy, there can be only one punishment.' He unslung the old Lee Enfield rifle and rammed home a round, the bolt making an audible click. 'Is that understood?'

The two men in question looked scared to death. Sher Dil slung the rifle over his shoulder again. 'Right, Corporal Nadin. Prepare to move out.'

Nadin and the three privates hurried away and Father Kerrigan heaved a sigh of relief. 'You almost frightened me.'

'A bad business,' Hamid said. 'Once it starts, you can never tell where it's going to end.'

Sher Dil nodded. 'We've wasted enough time. Get what you need in the way of arms and so on and we'll move out.'

The refugees, a dozen women and five children, huddled together in the shadows at the back of the truck, clutching the pathetic bundles which contained their worldly possessions.

They sat there patiently, watching with no visible emotion as Drummond and Hamid looked through what was left of the general equipment the truck had been carrying. They found quilted parkas similar to the one Ahmed was wearing and Drummond discarded his flying boots, still saturated from their immersion in the river, and helped himself to a pair of heavy mountain warfare combat boots. He pulled on waterproof mittens and jumped to the ground.

Hamid was at the second truck with Sher Dil. The colonel had discovered a case of sub-machine guns and had broken it open.

'These are very good,' he said with a grin. 'A gift from Moscow. One of the happier results of adopting a policy of strict neutrality.'

He prised open a box of ammunition and another of grenades and turned as Corporal Nadin approached. 'Bring the others, I'm going to issue automatic weapons.'

Nadin called and a moment later, Ahmed and the third driver, a tall Bengali named Amal, hurried out of the mist.

'Yussuf and Piroo – where are they?'

The two men glanced at each other uncertainly and Nadin ran along the line of trucks. He was back in a moment. 'They have gone, Colonel.'

Sher Dil grabbed Ahmed by the front of his parka. 'Did you see them go, you rogue?'

Ahmed raised his hands, palms outwards. 'On my father's grave, Colonel. They were here only five minutes ago. I was talking to them.'

'What about?'

'They were very angry with the Colonel. They said that the Chinese would catch us all. That we would never reach India.' He shrugged. 'They didn't want to stay.'

Sher Dill cursed and Hamid shook his head. 'We're better off without them. There's no problem. We've enough drivers between us. I can take a truck myself.'

Sher Dil nodded. 'Very well. I'll go first with Corporal Nadin. You follow in the supplies truck, Major. Father Kerrigan, Miss Tate and the young Khan can travel with you.'

'What about me?' Drummond said.

'You can bring up the rear with Ahmed. Mr Brackenhurst can travel with Amal in the third truck with the refugees. As soon as everyone's in, we'll move out.'

As they broke away, Drummond heard his name and saw Janet leaning over the tail-board of the second truck.

He climbed up beside her quickly. 'Anything wrong?'

'No, what's been happening?'

'A couple of drivers have deserted, but, there's nothing to worry about. How's Kerim?'

'Asleep at the moment. We've made him as comfortable as possible.'

Boxes had been moved away from the far end creating an alcove in which the young Khan lay covered with blankets, his face very pale against the white bandage. Janet leaned down to straighten a blanket and when she stood up again, Drummond took her hands.

'Are you worried?'

She shook her head. 'Nothing's really sunk in yet. I can't quite believe it's happening.'

His hands tightened, pulling her close and he kissed her. 'Not even this?'

She looked up at him, her eyes dark and serious, and then she smiled and touched his face gently. She didn't speak, there was no need and they kissed again.

'I'll see you later,' he said and left her there.

When he vaulted over the tailboard, Father Kerrigan was standing in the rain, a long cheroot jutting from his teeth beneath the shovel hat.

'And would it be all right if I got in now?' he demanded.

Drummond grinned and gave him a push up over the tailboard. 'Where did you get the smokes?'

'See Ahmed. He's been foraging amongst the supplies again.'

Drummond trudged through the mud to the rear truck. When he climbed up into the cab, he found Ahmed sitting behind the wheel wreathed in tobacco smoke.

The Afridi grinned and took a carton from the dash-board. 'Cheroots, sahib, very strong. Specially made for Indian Army.'

'The way I feel, I could smoke any damned thing,' Drummond said.

He lit one, coughing as the smoke caught at the back of his throat, the door was pulled open and Sher Dil appeared.

'The next village is Hasa which is a good ninety miles further on.'

'We haven't a hope of getting that far before darkness in weather like this,' Drummond said.

Sher Dil nodded. 'If we can make forty

miles I'll be satisfied. We'll camp at the side of the road and push on at dawn.'

He slammed the door, and Ahmed pressed the starter. After several moments and a liberal use of the choke, the engine rumbled into life. The truck in front of them lurched forward and he eased off the handbrake and followed.

There was a warm smell of petrol and oil in the cab and rain splashed against the windscreen. Suddenly, Drummond had that same feeling of temporary security and safety he'd known in the herdsman's hut after they'd got across the river. He leaned back in his seat, laid the sub-machine gun across his knees and started to clean the grease from it with a piece of rag.

As the sound of the truck engines faded into the mist and rain, Piroo and Yussuf scrambled up from the stream bed and stood in the rain listening.

As the last echo died away, Piroo nodded

in satisfaction. 'Good, they have gone. Sher Dil was very angry.'

'No matter,' Yussuf replied. 'His day is done.' He looked up at the smoke rising from the headman's house. 'There is still a fire on the hearth. We will stay here for the night. We can move on in the morning.'

They went up the steps to the verandah, opened the heavy door and went inside leaving the street empty again. Rain hammered into the mud, mist enfolded the silent houses and the village waited as night fell.

The truck rocked violently as it ground its way along the muddy, pot-holed road and Drummond leaned forward, straining his eyes into the swirling mist.

The truck in front stopped suddenly and Ahmed stamped his foot on the brake. Drummond opened the door, sub-machine gun ready, and Sher Dil appeared.

'We've bogged down. You'll have to lend a hand.'

Drummond and Ahmed tramped through the mud to the front truck. Its offside wheel was deep in a waterfilled pothole and Nadin and Hamid were already busy with spades.

It took twenty minutes of hard work on the part of everybody to get it moving. When Drummond climbed back into his seat, he was plastered with mud to the knees and his fine new parka looked as if it had been through a hard campaign. Half an hour later, the whole performance had to be repeated.

When he settled himself back in his seat for the second time he was past feeling anything. His feet were numb, his hands raw and bleeding from handling the rocks and stones which had gone to fill the potholes.

Visibility was bad now and he began to feel very tired as he strained his eyes through the gathering gloom. The front truck's horn sounded once and as the convoy slowed, he was aware of scattered pine trees on the left.

Ahmed turned off the road and followed the dimly-seen tail of the truck in front and there was a sudden stillness as all engines were cut.

10

Nightwatch

The camp site Sher Dil had chosen was a rocky flat, thinly scattered with pine trees that gave them some sort of a screen from the road.

When Drummond walked up the line, Hamid, Sher Dil and Father Kerrigan were standing at the rear of the second truck talking in low voices. Janet leaned over the tailboard.

'We've decided we don't need to worry too much about a blackout in this mist,' Sher Dil said. 'We'll set up one of the oil stoves in the back of the supply truck. Miss Tate can cook in there away from the rain. The refugees can do the same. There's plenty of food to go round.'

'A good hot meal should go a long way towards raising everyone's spirits,' Father Kerrigan said.

Drummond nodded. 'What about the boy?'

'He'll be all right. I've kept him under strong sedation so far.'

'What about sleeping arrangements?' Hamid asked.

'In the trucks. We'll need a guard, of course. Two at a time. One here, the other at the roadside. I'll work out a rota after we've eaten.'

Sher Dil moved away and Father Kerrigan smiled up at Janet. 'Hand me my bag, my dear. I'd better have a look at Brackenhurst.'

'I'll go with you,' Hamid said.

They walked away together and Drummond called to Ahmed and helped Janet down. The little Afridi arrived on the run. 'Yes, sahib?'

'Miss Tate's going to cook a meal for us in the back of the supply truck,' Drummond said. 'Get the spirit stove going for her and open a few tins. If you don't do everything she tells you at the double, I'll cut your throat.'

Ahmed grinned at Janet. 'The sahib has a

kind heart, memsahib. He could never do such a terrible thing. You come with me. I will see to everything.'

They moved away and Drummond went after Father Kerrigan and Hamid. He found them in the back of the third truck with the refugees. An engine inspection light had been rigged up to illuminate the interior. Brackenhurst sat on an ammunition box, stripped to the waist, as Father Kerrigan carefully peeled layers of bandage away from his left arm and the women and children watched solemnly.

He looked pale and drawn and every so often glanced furtively at Hamid who watched calmly. The priest removed the final bandage, examined the arm and nodded.

'Nothing like as bad as I thought at first. You'll be fine in a day or two.'

'It hurts like hell,' Brackenhurst said.

'What a shame.' Drummond pulled himself up and looked over the tailboard. 'Don't you think that's a shame, Ali?'

'Undoubtedly,' Hamid replied calmly. 'You must rest, Tony. We wouldn't want anything to happen to you.'

Brackenhurst threw them both a glance of pure hatred and Drummond dropped to the ground and moved back to the supply truck. He could smell cooking, and a sudden, hollow ache told him how long it had been since he'd eaten. When he climbed inside, Janet and Ahmed crouched over the stove and Sher Dil sat on a packing case, the map across his knees.

'You look worried.'

'I'm thinking about tomorrow. We swing very close to the river again. If the Chinese have moved fast along the other side in their troop carriers and get a few patrols across, we could run into trouble. The bridge across the Sokim Ravine, for example – if that was destroyed, we would be on foot.'

'We can worry about that in the morning,' Drummond said. 'I'm only interested in one thing at the moment – food.'

Ahmed passed plates of stewed meat and beans across, and as they started to eat Father Kerrigan climbed up, followed by Hamid.

'Will you see that Mr Brackenhurst gets something to eat,' the priest said to Ahmed and frowned at Drummond. 'Weren't you a little hard on him back there? Any man's nerve can go for a while in a situation like this.'

'He never had any in the first place,' Drummond said flatly.

The old priest frowned, glancing from one to the other, aware that there was something here that he did not understand. Drummond jumped over the tailboard and went round to the cab.

He sat in the warm darkness smoking and Ahmed brought him a mug of strong, scalding tea. A little later Sher Dil opened the door.

'I've worked out a guard rota. I'd like you to take over from Amal up on the road at ten. You'll also be on guard down here for an hour from 4 a.m. I want everyone up at five. We've got a long day ahead of us.'

He vanished into the darkness and Drummond pulled the fur-lined hood of his parka over his head. Ten o'clock. That left time for a couple of hours sleep. He settled into the corner and closed his eyes.

* * *

He was running down a long, dark road and somewhere ahead of him was Janet. She was calling to him and he knew that something terrible was close behind. He ran faster and then the surface of the road changed to mud and his feet began to stick in it, ankle-deep. Clouds of rain blew across his path, blotting her from sight, and only the sound of her voice told him she was still there. It became fainter and fainter and then he felt terribly afraid and the thing behind him, the nameless evil that made him so afraid, grabbed him violently by the shoulder.

He awakened suddenly to the cold night and realised that somebody was shaking him. He groaned and sat up. Ahmed said from the darkness. 'You were having a bad dream, I think, sahib.'

'Is it time?'

'Yes, sahib.'

Drummond breathed deeply a couple of times to steady himself, then pulled on his mittens. He picked up his sub-machine gun, opened the door and jumped down into the mud.

The rain rushed steadily through the darkness and the mist still blanketed the wet ground as he moved through the pine trees to the road.

After a while, he paused and called in a low voice, 'Amal, where are you?'

The Bengali moved out of the night to join him. 'Drummond sahib?'

'Anything doing?'

'Nothing, only the rain and yet more rain. Soon it will be snow. I have known it happen before this early in the year.'

'Let's hope not,' Drummond said and the Bengali faded into the darkness.

He found a fallen tree and sat on it, arms folded, sub-machine gun across his knee, but the cold ate into his bones and from time to time he stood up and walked around a little, stamping his feet to restore the circulation. Finally, with a complete disregard for caution, he lit a cheroot. It tasted terrible, but the glowing end somehow comforted him. When he had finished it, he lit another.

He became aware of the noise very gradually. He straightened up and listened carefully.

He could hear the sound of feet squelching through the mud from the direction of the camp. There was silence for a moment as if the person approaching was momentarily at a loss and then the steps sounded again, this time much more cautiously.

Very carefully, Drummond placed his still burning cheroot in a branch of the fallen tree, then slipped quietly into the darkness.

He worked his way round in a wide circle until he was certain he was behind the intruder and then moved forward. The dim bulk of a man appeared from the gloom, and glowing faintly through the dark beyond him was the cheroot.

It was the stillness of the man that decided Drummond, that and the slight, ominous rattle of a gun sling as he eased from one foot to the other, still peering towards the glowing cheroot. Drummond took a pace forward, tapped him on the shoulder and punched him in the stomach as he turned round.

He lay moaning on the ground and Drummond struck a match. It was Brackenhurst, one of the Russian sub-machine guns lying in the mud beside him. The match hissed and was extinguished by the rain.

After a while, Brackenhurst groaned and sat up. 'What happened?' His voice quavered and he sounded sick.

'You shouldn't go creeping around in the dark like that,' Drummond said. 'People might get the wrong idea.'

'I wanted a word with you, that's all,' Brackenhurst said. 'Away from the others. I wanted to explain about what happened at Sadar. When the roof started to come down, I panicked. Didn't know what was happening. I got to the Land Rover and when no one else followed, I thought you'd all bought it.'

Which was a straight lie, but Drummond let it go. 'That's all right. These things happen.'

Brackenhurst hesitated. 'Have you told anyone else?'

Drummond shook his head. 'Only Hamid and I know and we've more important things to worry about.' He stood there, calm and somehow uncompromising in the darkness. 'You'd better get some sleep. You're going to need it.' He picked up the sub-machine gun and held it out. 'Better take this with you.'

Brackenhurst stumbled away without speaking and Drummond went back to his tree. Half an hour later, Sher Dil relieved him. 'Anything happened?'

Drummond shook his head. 'No, everything quiet up here,' he replied and trudged through the mud back to the camp.

He climbed into the back of the truck and lay down, hitching a blanket over his shoulders. He was cold, numb all over and yet he wasn't miserable. He was long past that point.

He came awake slowly, yawned and turned on one side. Janet crouched over the oil stove, waiting for the kettle to boil, her face half in shadow in the subdued glow.

'What time is it?' he asked softly.

She glanced at her watch. 'Just after three. I couldn't sleep.'

She made tea in two tin mugs and handed him one and they sat in companionable silence in the glow of the stove. After a while, he said gently, 'What is it, Janet? Are you afraid?'

'I think I am,' she said simply. 'Even Vietnam didn't prepare me for anything like this. Do you think we'll get out?'

He was tempted to answer with a false assurance and then looked into that calm, grave face and knew that he couldn't. 'I'm not sure. As Sher Dil says, if the Chinese have moved fast along the other side, they could be ahead of us. They're bound to find boats at Huma or one of the other riverside villages. They could put men across to cut the road with no trouble.'

'This bridge up ahead that Sher Dil mentioned. Do you think there may be trouble there?'

'Trouble is where you find it. There's never any sense in worrying too much in advance.' He smiled. 'What will you do when all this is over?'

'Carry on to Chicago with Kerim, I suppose. That still stands whatever else may be changed. I'm due three months leave anyway.'

'And afterwards?'

'I'm not sure. I go wherever the Society sends me.'

'Isn't it time you thought about settling down?'

'Is that a straight offer?'

He shook his head. 'I could offer you money, Janet, enough and to spare. But take a look at the debit side. I'm forty years old, a beat-up ex-Navy flyer who's seen too much of hot countries and strange cities, had enough of flying to places where no one else will go. I want to rest my head somewhere for a while. That doesn't sound like much of a catch to me.'

'I know one thing,' she said quietly. 'If we don't try, we'll regret it for the rest of our lives.'

He sat staring down at the light of the stove, her hand in his, sighed and got to his feet. 'I'll get a little air. I've got some thinking to do.'

Janet sat there in the darkness, and after a while Hamid climbed over the tailboard. He helped himself to tea and squatted on the other side of the stove from her.

'I should wake Jack. He's supposed to take over from me at four.'

'That's all right. He was here. He's just gone for some air.'

'Trouble?' Hamid said.

She shrugged. 'Four o'clock in the morning talk, that's all. He's just decided he's too old for me.'

Hamid nodded. 'He's tired, that's all.' He hesitated and then decided to carry on. 'Jack isn't aware that I know this, Janet, but for the past five years at least, he's been working for British Intelligence, mainly flying illegal reconnaissance flights across the borders of what might be termed the less friendly powers.'

The breath went out of her in a long sigh. 'You're sure about this?'

'Oh, yes, the information comes to me from friends in Indian Army Intelligence. A long time for a man to live on his nerves.'

'Which explains a great many things.'

'Last year he crashed in the Borneo jungle and was badly wounded. They nearly got him that time and the Indonesians do not care for the British these days. His fate would hardly have been a pleasant one.'

'Is that when he got that terrible scar on his face?'

He nodded and leaned across, his face grave in the diffused light of the stove. 'He's a good man, Janet, but he's had enough. Take him home, wherever that turns out to be.'

There was an obvious change in her, she seemed confident, assured, smiled suddenly and squeezed his hand. 'I will, Ali, I will.' She got to her feet. 'I'd better check on Kerim.'

Hamid poured himself some more tea, feeling strangely sad, and after a while Drummond climbed over the tailboard and joined him.

'Where's Janet?'

'Gone to have a look at Kerim. Who's on duty at the road?'

'Ahmed, I think.' Drummond hesitated and then went on, 'Brackenhurst turned up when I was doing my hour up there last night.'

'What did he want?'

'I'm not sure. I rather got the impression he was toying with the idea of killing me, but basically he just wanted to make sure that we hadn't told the others what really happened at Sadar.'

He explained what had taken place. When he had finished, Hamid nodded slowly, a frown on his face. 'Of course, he could argue that he was carrying the gun merely as precaution in case of trouble. He could never have used it. Far too noisy. The knife is the weapon for darkness, Jack.'

'I'm not sure that he's rational enough to look at things in that way any more,' Drummond said. 'He's badly scared, and he's certainly never possessed the kind of cold-blooded guts it takes to go after a man with a knife.'

'We'll have to watch him from now on, that's all,' Hamid shivered suddenly. 'I don't like it at this time in the morning, Jack. Makes me think of other dawnings, other places and a lot of good men dead.' He laughed in a peculiar fashion. 'I must be getting old.'

'Aren't we all?' Drummond said.

He got to his feet and moved to the tailboard. It was already dawn, a grey light seeping through the mist. The heavy rain lanced into the ground and he stared out at it morosely, wondering what the day would bring.

Wrapped in a sheepskin on the floor in front of the fire in the headman's house at Bandong, Piroo was awakened suddenly by a savage kick in the side. He sat up with a start, aware, as if in a dream, of faces staring down at him, the shining Burp guns, the red stars in the peaked caps.

Somewhere, Yussuf cried aloud, running for the door. A foot tripped him and a rifle butt thudded savagely against the back of his skull, cracking the bone.

Piroo was dragged to his feet, gibbering with fear and then a sharp voice cut across the noise and confusion and there was silence.

Colonel Cheung paused in the doorway, the fur collar of his greatcoat pulled up around his neck, the face beneath the fur hat lined and drawn with fatigue.

There had been considerable delay in

crossing the river at Kama. For one thing, the shallows had been deeper than usual owing to the heavy rain and one of the troop carriers had bogged down. They had wasted several hours in trying to salvage it. It had been almost dark when he had finally decided to push on with the remaining vehicle and a dozen men.

He had kept on the move for most of the night, often at no more than ten miles an hour in the appalling conditions, on several occasions almost losing the vehicle, but there was always the hope that Father Kerrigan and his party might be at Bandong. It was an obvious stopping place. When they had reached the village, he had sent the sergeant and ten men in on foot, giving them five minutes' start before following in the troop carrier.

'What's going on here?' he demanded.

The sergeant, a small, hard-faced Cantonese named Ng, hurried forward. 'The village is empty, Colonel, except for these two. Deserters from the look of them.'

'Deserters?' Cheung's face changed, went

pale with excitement as he pushed his men aside and examined Piroo. 'Who are you?' he demanded in Urdu. 'One of Colonel Sher Dil's men? Did you escape across the river?'

'No, sahib,' Piroo said. 'I was with the supply convoy.'

'The convoy was here?' Cheung said. 'Where is it now?'

'Gone, sahib, to India with Colonel Sher Dil and the young Khan. They are hoping to reach the border.'

'Sher Dil was here?' Cheung said in amazement.

'Oh, yes, sahib,' Piroo babbled. 'Also Major Hamid and Drummond sahib. They all crossed the river from Sadar.'

'When did they leave?'

'Yesterday, two hours before dark. They intended to camp somewhere along the road when night fell. I heard the colonel say so.'

Cheung laughed excitedly, slapping his leather swagger stick across his gloved palm. 'Get the men together at once, Sergeant. We're going on.'

He turned to the door and Sergeant Ng

said quickly, 'What shall I do with this one, sir?'

Cheung looked at Piroo with something close to affection. 'Let him go free, he has served us well.'

He went out and Piroo, to whom the interchange in Chinese had been completely unintelligible, turned eagerly to Sergeant Ng.

A strange man, the colonel, the sergeant told himself. Full of wild fancies, but a good officer for all that. He nodded to one of his men who grabbed Piroo suddenly, clamping a hand over his mouth.

Piroo saw the knife coming up, felt a coldness streaking under the ribs to the heart and plunged into darkness. They left him there by the fire, and a moment later the troop carrier moved away, its tracks scattering mud from the street across the walls of the houses.

11

The Bridge at Sokim

'I can see the bridge,' Sher Dil said, 'and it's still standing.'

'Thank God for that.' Drummond took the binoculars and focussed them quickly. 'There certainly doesn't seem to be anybody about.'

'And no cover for an ambush,' Hamid said. 'We'd better cross now while the going's good.'

They dispersed to the trucks and Drummond ploughed through the mud and heaved himself up beside Ahmed, glad to be back inside. It had rained without pause all morning, turning the road into a rutted quagmire through which they had progressed at little more than fifteen miles per hour.

They went over the hill and the road dropped steeply towards the great ravine which cut its way through the heart of the mountains. Ahmed selected bottom gear and followed Sher Dil cautiously.

The bridge was a narrow spindly thing, fit only for one-way traffic. As the road levelled off to approach it, the other trucks slowed to a halt and Ahmed braked quickly.

'I'll see what's happening,' Drummond said and jumped down.

Sher Dil leaned over the parapet examining the web of rusty steel girders. He turned as Drummond approached.

'It would take the Chinese a long time to construct another. A demolition expert's dream.'

'Thinking of doing it yourself?'

'I don't see why not. It wouldn't take long. We'll cross over first, though.'

As Drummond went back to his own vehicle, Hamid leaned out of the cab of the supply truck. 'What was he up to?'

'He wants to stop and blow up the bridge. What do you think?'

'An excellent idea. It would block the road for months.'

'Don't you think it might advertise our presence?'

'I can't see that it makes much difference. If there's anyone up ahead, they'll still be there whether we blow the bridge or not.'

Drummond climbed up beside Ahmed and the truck lurched forward and started the slow ascent on the other side. When they pulled over the hill, they saw that Sher Dil had stopped a little way up the road. They joined him as Amal and Brackenhurst drove up behind.

Brackenhurst came forward, his face white and strained. 'Why are we stopping?'

'I've decided to destroy the bridge before moving on,' Sher Dil said.

Father Kerrigan climbed down to join them and Janet stayed in the cab, an arm around young Kerim who was now sitting beside her.

'For God's sake,' Brackenhurst said. 'Haven't we lost enough time?'

'If we blow the bridge, the Chinese will lose even more,' Sher Dil said patiently. 'We'll use the contents of my truck, grenades, ammunition and some demolition charges. You can all help. We'll unload the stuff more quickly that way.' He turned to Father Kerrigan. 'You stay with Miss Tate and the Khan. We shan't be long.'

He took the wheel himself on the journey back down the hill. When they reached the bridge, he turned and reversed as far as the centre. Drummond climbed into the back with Hamid and handed the boxes out to the others. They worked fast and each time he passed a box down to Brackenhurst, he noticed that he was sweating.

'That's about it,' Sher Dil said at last as he surveyed the boxes stacked across the bridge. 'When that lot goes up, they'll hear it in Sadar.'

'What now?' Hamid said.

'I'll set the fuses myself. Corporal Nadin and Amal can stay and help me. The rest of you better get back up the hill. You'll have to walk. We'll need the truck to make a quick exit before the big bang.'

Brackenhurst was already hurrying back

across the bridge and Nadin stood rooted to the spot, dismay in his eyes. Sher Dil tossed him a coil of fuse wire which the Indian almost dropped.

'Pull yourself together, Corporal,' the colonel snapped. 'The sooner we get this set up, the sooner we can get out.'

At the top of the hill, Drummond turned and looked down. The bridge and the truck seemed like toys and the whole scene had an unreal, fake look.

Hamid came up the hill carrying Sher Dil's binoculars. He sat on a boulder and adjusted them until the truck and the bridge jumped sharply into focus.

'How are they doing?' Drummond asked.

'He's laying the charges. I must say Nadin doesn't look too happy. Neither does Amal.'

'They're both scared to death. I think that's why Sher Dil made them stay.'

Below on the bridge, they worked rapidly. Nadin running the fuse wire to the far end. He walked back towards the colonel, paused and pointed dramatically. His cry rose thin and clear in the rain.

As a Chinese troop carrier came over the crest of the hill on the other side of the ravine, Hamid focussed the binoculars quickly and the face of the officer standing upright beside the driver jumped out to meet him.

'It's Cheung.'

As the troop carrier started the descent, Drummond said, 'They haven't got time to blow the bridge now. We'd better get moving.'

'In the troop carrier, they would catch us within five minutes,' Hamid said calmly. 'Sher Dil knows that. He will blow the bridge. He will do it for the young Khan.'

Drummond turned to watch the drama being enacted below, saw Sher Dil take a step towards the charges and knew with cold certainty that Hamid was right, that he intended to blow the bridge even if it meant going up with it.

Amal seemed rooted to the ground, but Nadin rushed at him in complete panic, clawing at his shoulder. Sher Dil knocked him down with a blow of his fist and turned again to the charges. Nadin scrambled to his feet, wrenched the truck's spade from its fastenings near the door and struck Sher Dil on the head with all his force.

He turned and jumped into the cab. The truck moved forward and stalled and Amal, seizing his chance, scrambled over the tail-board. Sher Dil managed to reach his knees. He took a grenade from one of his pockets, pulled out the pin and tossed it towards the stacked boxes.

At that moment, the truck lurched forward. It had moved perhaps ten yards when the centre of the bridge erupted in a cloud of smoke. Pieces of stone girder lifted skywards as a series of violent explosions sounded one after the other and then the entire middle section of the bridge fell in, the truck slipped back into the gaping chasm and disappeared.

The troop carrier had slewed to a halt on the other side and now its heavy machine gun opened up, firing blindly through the pall of smoke, bullets ripping up earth and stone on the hilltop beside the trucks.

Brackenhurst was already behind the wheel of Amal's truck, lurching down the road to a chorus of terrified screaming from the women and children in the back.

There was no time to talk. Hamid scrambled up behind Father Kerrigan and drove away quickly and Drummond and Ahmed followed in the supply truck. For a heart-stopping moment, bullets ripped through the canvas hood and then they were over the hill.

Ten minutes later, Hamid sounded his horn, Brackenhurst slowed, turning in to the side of the road and Ahmed and Drummond pulled in ahead of him.

Drummond jumped down to join the Pathan and Brackenhurst stumbled towards them, his eyes wild. 'What do we do now, for Christ's sake?'

Hamid ignored him and held up Sher Dil's map. 'He left this in the cab, that's one good thing.' They leaned over it and he nodded. 'I thought so. There's another village fifteen miles further on and the border, fifty miles beyond that.'

'One thing's certain,' Drummond said. 'Cheung can't hope to catch us now.'

Hamid nodded. 'As long as there's no one

waiting at the next village we should be all right.'

'What about ditching a couple of the trucks?'

Hamid shook his head. 'If anything went wrong with the truck we were travelling in, we really would be in trouble. Another thing – three trucks give us more of a show of strength than one. That could be useful if we run into any small patrols.'

'What about the women?' Brackenhurst demanded. 'Don't you think it's time we left them?'

'For the Chinese to get their hands on? Even for you, that isn't such a sparkling idea. Get back to your truck and take up third position.'

The contempt in Hamid's voice was obvious to them all. Brackenhurst turned as if he had been struck and stumbled away.

'For heaven's sake, Major,' Father Kerrigan said, leaning out of the cab of Hamid's truck. 'The man's at the end of his tether, can't you see that?'

'Which means he's got to be driven, Father. It's the only way, I'm afraid.' Hamid turned to Drummond. 'You and Ahmed take the lead in the supply truck, Jack, and I'll follow. If you do run into trouble, try to block the road with your truck. That'll give the rest of us a chance to turn round. If you run fast enough, you should be in time for a lift.'

'I hope so,' Drummond said.

He went round the front of Hamid's truck and waved to Janet. She waved back and so did the boy, the first real sign of life Drummond had noticed from him.

He climbed up beside Ahmed and they drove away. The mist had lifted even more which wasn't too healthy and the rain sluiced down relentlessly. It was bitterly cold and there was a bad taste in his mouth. He ran a hand over the stubble on his chin and leaned back in the seat watching the road, a submachine gun in his lap.

They had been driving for no more than twenty minutes when Hamid sounded his horn a couple of times. Ahmed turned in to the side of the road and Hamid pulled in ahead.

He jumped down and came to meet them. 'Brackenhurst doesn't seem to be following.'

'I wonder what the stupid bastard's up to now?' Drummond said.

'As far as I'm concerned, I'd leave him to stew, but we've the women to think of.'

Drummond nodded. 'You wait here. We'll run back in the supply truck. It's always possible that he's just broken down. I can't understand how the things have kept going this long.'

Ahmed reversed and drove back along the road. Within five minutes they saw the truck parked at the edge, the women and Brackenhurst standing beside it.

He drove past, turned in a tight circle and parked a few yards away. Drummond jumped down and walked back. Brackenhurst was smiling nervously, relief on his face.

'Thank God you came. I knew you would.'

'What happened?' Drummond asked.

'It's the brakes. They're hydraulic on this truck. They've stopped working. There must be a leak.'

'That's all we needed.'

'We'll have to leave the women now,' Brackenhurst said.

'Take a look underneath, Ahmed,' Drummond said and he climbed into the cab of Brackenhurst's truck.

He pumped the brake pedal up and down several times, but there was no answering pressure. At that moment, Ahmed called to him. He jumped down, pushed his way through the women who had crowded silently around and crawled under the truck.

'See, sahib,' Ahmed said grimly. 'The pipe has been deliberately fractured.'

As Drummond started to examine it, the engine of the supply truck burst into life. He scrambled out frantically, but he was too late. As he shoved the women out of the way, Brackenhurst accelerated. For a little while there was the sound of the engine and then that too died away and there was silence.

Ahmed moved to his side. 'I think there must be a special place in hell reserved for Mr Brackenhurst. What do we do now, sahib?'

'Go after him, what else can we do?'

'Without brakes, sahib?'

'It wouldn't be the first time. I'll take the wheel.'

He turned wearily to the women as a small child started to cry. Its mother hushed it and there was silence again as they waited, stolid and patient.

'Get in!' he said. 'Go on, all of you! Get back in!'

God knows what Brackenhurst would do when he reached Hamid and the others, probably keep right on going. And there was no means of knowing what Hamid might do. Best to try and catch up with him as soon as possible. They could unload the ammunition and transfer the women. If he drove carefully and used the gears, he could manage without the foot brake.

He climbed behind the wheel, eased off the handbrake and took the truck slowly forward. After a while, he gained more confidence, moved into top gear and put on speed. Within five minutes he reached the place where he and Hamid had stopped and rolled slowly to a halt. He could see the tyre marks at the side of the road, and an oil splash, but that was all.

Which wasn't good and he took the truck forward again grimly. There was hail mixed in with the rain now, building up against the windscreen and the wipers were having difficulty in handling it. After half an hour, the road started to slope down.

He changed to a low gear and proceeded more cautiously. The valley widened until it stretched away into the rain for about half a mile, lifting into the mountains that could be seen dimly on his left. The road dropped even more abruptly and peering through the misty windscreen, he saw a small bridge.

He crawled the rest of the way down the hill in bottom gear. The bridge consisted of a flat surface of planks crossing what would otherwise have been a deep ford. There was still no sign of either of the other trucks and he drove across and kept on moving.

The road started to lift steeply, hugging the side of the mountain which now towered above his head and he began to sweat a little. The truck churned steadily upwards through the mud, Drummond gripping the wheel tightly, an expression of utter concentration on his face. He rounded a curve and came to the crest of a hill and the road dropped steeply into the valley below. He leaned across quickly and looked out. There was no fence, only the crumbling, rain-soaked edge and two hundred feet of steeply sloping mountainside.

The truck began the descent, skidding occasionally with a sickening lurch. Drummond was trembling, and beside him Ahmed's face was wet with sweat. The truck lurched again as he negotiated a corner and then the wheels skidded on the shaky surface and slid forward for about fifteen yards. He turned into a skid and then out of it, and by a miracle regained control.

The sweat soaked through his shirt, ran from his forehead into his eyes and he took the truck forward again, hail rattling against the windscreen in a flurry of wind.

The road curved around a great outcrop of black rock and he followed it, hugging the side, turning the corner to where an apron of brown and white water flooded the road, rushing down from the mountain above and cascading into space.

As he started across, the front wheels dipped and the surface of the road dissolved beneath him, washed out in a great sliding scoop, and the truck slewed towards the edge.

For a moment, it seemed to halt and Drummond tugged frantically at the hand-brake, but it was not enough. The truck lurched and one of the front wheels dipped over the edge.

'Jump for it!' he cried to Ahmed.

He got the door open and went out head first, landing on his hands and knees, slipping in the thick mud as the truck slid past him and went over the edge.

It hung there for a split second, and Ahmed, whose door had jammed, got it open a second too late. The truck went over with a chorus of screams from the wretches imprisoned inside. There were three terrible, metal-wrenching crashes as it bounced its way down the valley, a brief moment of silence and then a tremendous explosion.

Drummond moved cautiously to the side and looked down at the bright, orange tongue of flames and turned away, his body heaving in a great, uncontrollable spasm that emptied his stomach.

He leaned against the rock for a while and then scrambled across the washed-out portion of the road and walked on into the rain.

For half a mile the road dropped steeply into the valley and he caught a brief glimpse of the river, winding through the mist below. The rain became colder and darkness started to fall.

There was only one way to go, though God knew where it would take him. He wasn't even armed. His sub-machine gun had gone over the edge with the truck.

Something brushed his face. He raised a gloved hand and saw that it was covered with large melting snowflakes. He looked up and all around him, snow fell intermingled with the rain.

From further along the road came the rattle of small arms fire and he paused for a moment, the snow gently covering him. Who was it? Hamid or Brackenhurst? But there was no means of knowing and he started to walk again.

Darkness increased and the snow gradually took the place of the rain until it was falling all about him, covering the muddy road with a white mantle. Again there came the rattle of small arms fire, much nearer this time.

The situation was beginning to look desperate. If he stayed on the road, he was bound to run into trouble sooner or later to judge from the sound of that firing. Without shelter, he would freeze to death on a night like this.

Trees had now begun to cover the valley bottom and he moved into their shelter and stumbled along, parallel with the road, his gloved hands tucked into his armpits against the intense cold.

Somewhere up the road, there was the clatter of a hoof against stone, and a horse snickered softly. Drummond dropped behind a tree and waited.

There was a soft drubbing of hooves muffled by snow and half a dozen horsemen cantered by. They wore the typical rough sheepskin coat of the hillmen, but the red stars in the peaked caps, the Burp guns slung across their backs, told him what they were.

'What do I do now?' he said softly as the hoofbeats faded into the night.

There was a quiet chuckle almost in his ear and Ali Hamid said, 'Exactly what I was wondering.'

12

The Long Night

'When I first heard you coming, I thought it meant trouble. I was about to become most unpleasant.' Hamid smiled, his teeth gleaming in the darkness. 'A good thing those soldiers rode by. The moment you took cover, I knew you were on the right side.'

It was impossible to see his face in the darkness and Drummond reached out to touch him in sheer relief. 'Ali, you old bastard. What happened?'

'You tell me. We were waiting for you to come back with news of Brackenhurst and you went by as if half the Chinese Army was on your tail.'

'That was Brackenhurst, not me,' Drummond

said and explained quickly what had happened, including the loss of the truck.

There was a moment's silence when he finished and Hamid said softly, 'There was heavy firing up ahead, I think he may have paid the price already, Jack.'

'He couldn't,' Drummond said flatly. 'It's too heavy.'

'Perhaps, but since Sadar, I don't think he's really been responsible for his actions.'

'Where's the truck?'

'About fifty yards back in the woods. I decided to leave the road when we heard the firing up ahead. We obviously weren't going to get any further. I came back to make sure that the snow covered our tracks.'

'Judging by the soldiers, the next village is obviously in Chinese hands. What are we going to do?'

'I haven't the slightest idea. We'll discuss it in more comfortable surroundings. At least we should be safe here for the night.'

Drummond stumbled after him through the darkness and the truck loomed out of the night. 'Not exactly the Savoy,' Hamid said, 'but better than a snowdrift on a night like this. Careful, there are boxes all over the place. I dumped half the load.'

The canvas curtain at the back of the truck moved slightly, showing a chink of light and Father Kerrigan said softly, 'Major Hamid?'

'And guest,' Hamid said. 'The wanderer returns.'

Drummond followed him over the tailboard. He dropped the canvas curtain back into place and turned. As Hamid had said, half the vehicle's load had gone and the remaining boxes had been stacked so as to create a small enclosed alcove. An oil stove was set on a box in the centre throwing out life-giving warmth as well as a dim light.

Father Kerrigan murmured something, a hand on his shoulder, but Drummond had eyes only for Janet kneeling on the other side of the cooker next to the young Khan.

'Jack,' she said in a whisper. 'Jack?'

He moved close, dropped on one knee and took her hand. There were no words, none that would say the things he wanted to say and he touched it to his lips briefly.

'What happened?'

He told his story again in a few brief sentences. When he finished, there were tears in her eyes. 'Those women, those poor women and children. And Ahmed.'

'There was nothing I could do,' he said. 'Nothing.'

'I thought we'd never get round that mountain ourselves,' Hamid said.

In the silence that followed, Janet seemed to pull herself together and put the kettle on the stove. Father Kerrigan said slowly, 'Then the shooting we heard earlier? That must have been Brackenhurst?'

Hamid nodded. 'There were Chinese on horseback up on the road, That means they must be in the next village.'

'Are we safe here?'

'For tonight.'

'And in the morning?'

Hamid shrugged. 'I don't honestly know. Even if we could get the truck out of here again, which I doubt, there's nowhere to go. We'd never get through the village and we all know who's coming up behind.' He held out his hands to the stove. 'We've got shelter, food and warmth and that's a lot under the circumstances.'

'Beans,' Janet said. 'Beans and tea.'

'Sufficient unto the day, isn't that what the Bible says?'

She poured tea into two tin mugs and passed them to Hamid and Drummond. 'Those are all we've got. We'll have to share.'

Drummond took off his mittens and wrapped his frozen hands around the mug, conscious of the warmth and from the shadows opposite, Kerim watched him solemnly, swathed in blankets.

He smiled, showing even white teeth and Drummond smiled back at him. 'He's beginning to liven up.'

Father Kerrigan nodded. 'The natural resilience of the young, I suppose.'

Drummond sat there, staring into the fire, remembering many things. The city burning, the old Khan's eyes blazing into his as he exacted that final promise, Cheung's pale, handsome face. Strange how things turned out. They'd been very good friends, really. And what happened now?

He emptied the mug and passed it to Janet. 'Where's the map?'

Hamid produced it from a pocket of his parka. 'Any ideas?'

'Not at the moment. How far are we from the village?'

'Here.' Hamid pointed as Drummond spread the map on the floor. 'Perhaps five miles. It's called Chamdo. The border's about fifty miles on the other side.'

Drummond examined the map carefully and frowned. 'Where does this track go to running over the mountain from the village? There's a place up on top on the plateau. Ladong Gompa.'

'Ladong Gompa?' Father Kerrigan put in. 'But that's a monastery, a Buddhist monastery. There's a shrine in the next valley, very famous in the old days. Pilgrims used to cross over the mountain and stay overnight at the monastery. I believe that's why it was built in the first place. The old Khan told me about it once.'

Hamid examined the map and shook his head. 'That's eight or nine thousand feet up,

Jack, and the snow starting. Father Kerrigan and Janet could never get across.'

'But you could with the boy,' the priest said.

Drummond cut in quickly. 'We all could if we had horses.'

'Horses?' Hamid said with a frown. 'And where are we going to get horses?'

'As you said, the village is only five miles along the road. If we slipped in just before dawn, we shouldn't have too much trouble.'

'All of us?' Hamid said.

Drummond shook his head. 'Just you and I. The others can wait for us here. When we come back with the horses, we can cut up across the shoulder of the mountain and join the track above the village.'

'If we come back with the horses.'

'At least it gives us a chance.' Drummond shrugged. 'Can you think of anything better?'

Hamid shook his head slowly. 'That's the trouble, Jack, I can't. I don't suppose we have much choice.'

'Then I suggest we get some sleep. We're going to need it.'

Janet passed him a blanket and he wrapped himself in it and lay down next to Hamid and the old priest. Surprising how warm the stove had made the interior now. He looked across at Janet sitting against the boxes, head bowed, the young Khan sleeping in the hollow of her arm.

A wonderful girl. The shadows thrown by the stove on to the canvas hood moved in and out, now coming together, now separating. Just like people, he thought. Now they need each other, now they don't. Now they mingle with each other, now they go their own way.

He slept well in spite of the cold that crept into the truck during the night and found himself crushed between the old priest and Hamid. When he awakened he sat up and lit the stove. The bright flame reflected suddenly from Kerim's unbandaged eye and Drummond grinned at the little boy, huddled in the corner next to Janet.

He motioned him to silence and looked outside. It was that time just before dawn when things begin to take on shape again, to have definition. There wasn't anything like as much snow as he had expected. Quite obviously, it had stopped falling hours before.

He felt curiously refreshed and jumped down into the snow, enjoying the fresh air in his nostrils after the close atmosphere of the truck. As he stood there, the trees started to stand out with a sort of hard luminosity and he knew that dawn was not far away.

'Enjoying the morning air?' Janet said quietly from the truck.

He turned and smiled. 'I don't know if you could say that exactly.' He spread his hands in a vaguely French gesture. 'I feel funny this morning. Close to home, wherever that is, and yet I know I'm not.'

She reached down for his hand in the darkness and gripped it tightly. 'We'll get there, Jack, I know we will.'

'Well just go on believing that.' He grinned. 'Better put some tea on the stove and wake Ali. We haven't got much time.'

'No need.' Hamid looked out of the canvas

screen beside her and Janet moved back. 'What's the day like?'

'Could be worse. It can't have snowed for very long.'

'It'll be back, I can promise you that. We'd better get ready.'

Drummond climbed back into the truck and found Father Kerrigan crouched at the stove beside Janet, opening tins of beans.

'How do you feel?' Drummond asked.

Father Kerrigan smiled. 'The old bones are beginning to creak a little, but I'll manage.'

'One thing I didn't check last night. Can you both ride?'

Janet nodded. 'Since I was a child.'

The priest smiled. 'I should imagine you've been used to a rather more sedate mount than the local variety, my dear. Intractable brutes, I know from bitter experience.'

'I'll manage,' she said confidently. 'What about you, Jack?'

'I get by, but only just. Ali's your man. He's a Hazara. They spent about a thousand years galloping down into India and back again, usually with a woman across the saddle.'

Hamid grinned and broke open a case of Garrand automatic rifles and Drummond cleaned one quickly. He found a box of ammunition and slipped several spare clips into his pockets. Hamid primed half a dozen grenades and they took three each.

Janet called softly and they sat in a circle round the oil stove, drinking hot tea and eating beans. 'That's the last of the food,' she said. 'I can fill the big Thermos with hot tea before we leave, but after that, we've had it.'

Drummond finished his tea and handed her the mug. He glanced at Hamid. 'Ready?'

'As ready as I ever will be.'

Drummond shouldered his Garrand and dropped over the tailboard. When he turned to look up, Father Kerrigan and Janet were pale shadows in the darkness. 'We'll be back in a couple of hours,' he said, trying to sound confident and they moved off.

Hamid led the way through the trees, his boots crunching the crusted snow and as Drummond pushed frost-covered branches to one side with a gloved hand, a feeling of exhilaration took possession of him. It was going to be all right. It had to be. They'd come too far, suffered too much.

Hamid raised an arm and they halted. The road lay just in front of them. As they stood in silence looking at it, snow began to fall quietly in large, firm flakes.

A tall, black finger of rock lifted out of the gloom on the other side and he pointed to it. 'That's as good a marker as any. We might as well use the road, it'll be quicker, but keep your eyes open. I've a nasty feeling we've left a little late. It's getting lighter by the minute.'

And he was right. One by one, the trees seemed to step out of the darkness as they marched along the road. The muddy ruts were ice-bound and iron-hard with just enough snow covering them to make walking easy. They moved quickly, Hamid in the lead, Drummond behind him and keeping to the other side.

The snow was quite heavy now and reduced visibility considerably. There was that strange, absolute quiet that snow always brings and it affected Drummond powerfully so that for a while, he walked with his head bowed, oblivious to all possible danger, alone with his thoughts.

They had travelled for no more than half a mile when he was brought back to reality sharply by Hamid's low, urgent call. He was standing at the side of the road and Drummond hurried to join him.

The tail of a truck was sticking out from the trees at an unnatural angle perhaps twenty-five yards into the wood. They stood there for a moment, not speaking, both thinking the same thought and then Hamid led the way forward, following the snow-covered path the vehicle had made for itself.

It was the supply truck. Drummond brushed snow from the side of the vehicle and his glove snagged on rough edges. He regarded the bullet holes dispassionately.

'The thing's like a sieve. He must have run straight into trouble.'

He wrenched open the door, but the cab was empty and then Hamid called from the other side. Brackenhurst lay huddled under a tree, his face turned slightly, fingers frozen into tallons. There were three gaping holes in his chest.

They stood looking down at him and somewhere, a horse snorted softly. There was the jingle of harness and voices, soft on the morning air. Someone laughed and Hamid and Drummond slipped into the shelter of the trees.

At the end of the jagged lane the truck had made into the wood, two horsemen appeared, Chinese dressed in great sheepskin coats and peaked caps, guns slung across their backs. They reined in, looking down towards the truck and one of them laughed again.

Hamid handed his sub-machine gun to Drummond and said softly, 'Give me your rifle. We can't let them go on. They'll spot the other truck.'

Drummond gave him the Garrand and Hamid rested the barrel against the tree trunk in front of him. The horsemen had just started forward again when his first shot tumbled the lead man from the saddle. He screamed, turning on to his face in the snow. As both horses plunged in panic, the second rider fought to turn his mount. He was still trying when two bullets in the back lifted him from the saddle.

As Drummond and Hamid ran forward, one horse cantered away slowly, back towards the village. The other stood patiently beside the body of its rider. Hamid slung his rifle across his back, gathered the reins and vaulted into the crude sheepskin saddle.

'I'll catch the other one, Jack.'

He urged his mount forward and disappeared into the curtain of snow. Drummond checked the action of the sub-machine gun and waited impatiently. Somewhere in the distance, he seemed to hear a faint cry and then Hamid galloped back along the road, the reins of the second horse in his right hand.

'We'd better get moving. More horsemen back along the road. The bastards are out early this morning.'

Drummond slung the sub-machine gun across his back and took the reins. The horse moved away from him, rolling an eye and he pulled it back savagely and scrambled into the crude saddle.

Hamid urged his mount into a gallop and Drummond hung on grimly as his own horse followed. There was excited shouting somewhere to the rear, but no shooting and then the black finger of rock loomed out of the falling snow on their left and Hamid turned into the trees.

Father Kerrigan was standing anxiously beside the truck and Janet leaned over the tailboard as they dismounted. 'What happened?' the old priest said.

'Never mind now,' Drummond told him. 'Get the boy. We've got to get out of here.'

Janet handed Kerim down, slung a small military haversack over her back and followed him. Swathed in the grey army blankets, the boy looked like a little old woman and didn't seem to be in the least afraid, his large, dark eye taking in everything with interest.

Drummond gave Janet a leg up on to his horse and handed her the child. She settled him in front of her and took the reins.

'Across the road and up the hillside,' he said, 'and don't waste any time getting there.'

As Hamid helped Father Kerrigan into the saddle of the other horse, there was movement up on the road, voices called excitedly and then, quite suddenly, the sharp report of a rifle and a bullet thudded into the side of the truck.

Drummond unslung his sub-machine gun and gave Hamid a violent shove. 'Get out of it, Ali! I'll hold them.'

Hamid didn't argue. He vaulted up behind Father Kerrigan and smashed his clenched fist against the horse's hindquarters. It bounded forward into the trees and the other horse followed instinctively.

Drummond fired a quick burst through the brush towards the excited voices and someone cried out sharply. He ran from the shelter of the truck and dropped on one knee behind a tree.

He could hear the sound of his friends' progress somewhere to the left as Hamid took them away on a diagonal course, obviously intending to cross the road lower down.

A mounted soldier burst through the trees towards the truck, another behind him. Drummond loosed off a long burst that sent both men and horses down in a confused heap, turned and ran headlong through the trees, following the trail left in the snow by the others.

There was movement over to his right, dark shadows against the snow and he emptied the sub-machine gun in a great, sweeping arc and ran on.

As he emerged into a small clearing, a Chinese soldier ran out of the trees on his right. Drummond's sub-machine gun was empty. He dropped it and rushed straight at the Chinese at the same headlong pace.

The man was badly shaken. Instead of trying to aim his Burp gun, he raised it defensively. Drummond ducked under the flailing weapon, grabbed for the throat and lifted a knee into the man's crutch. As the Chinese sank into the snow, he tore the weapon from his grasp and ran on.

He was sobbing for air as he stumbled through the trees and scrambled up the little slope to the road. He slipped and fell to one knee. As he stood up and made to cross, he heard voices through the falling snow.

At least a dozen soldiers were running towards him, but these weren't mounted, they were on foot and wore normal quilted uniforms. And then he saw Cheung in his long greatcoat with the fur collar, mouth open in a soundless cry.

Drummond emptied the magazine in one continuous, clumsy burst that ripped up the surface of the road for twenty yards in front of the Chinese, ran across and started to scramble up the hillside.

He heard the roars of the men behind as they followed then a cry of alarm echoed by an explosion. A few seconds later, there was another. He kept on moving and fell on his face.

Hands dragged him to his feet and Hamid said, 'A good thing I had those grenades.'

Drummond leaned against him, feet splayed and fought for breath. 'The lot I ran into just now,' he said. 'Not soldiers from the village. Cheung and his men. They must have followed on foot from the bridge. Isn't the bastard ever going to give up?'

'I shouldn't imagine so,' Hamid slapped him on the shoulder. 'We'd better get moving. He'll need horses if he's going to follow and that means going to the village. It'll take time.' He grinned savagely. 'With any luck, one of my grenades may have finished him off. He could be lying down there in his own blood right now.'

And then the wind tore a hole in the curtain, and for a moment they saw the road below, the bodies sprawled in the snow, the living moving amongst them and one man who stood quite still, staring up at the mountain, the fur collar of his greatcoat framing the pale face.

'No such luck,' Drummond said with a shudder.

As the curtain swept back into place, he turned and followed Hamid upwards into the driving snow.

On the road, the carnage was absolute as Cheung turned to examine the dead and the dying. Only Sergeant Ng and three men were left on their feet, and then one of the soldiers from the village limped out of the wood clutching a bloody arm, his sheepskin wet with snow.

Cheung went to meet him, the sergeant at his side. 'You are from Chamdo, the next village?'

'Yes, Colonel.'

'How did you get there?'

'By boat from Huma. Two patrols crossed straight over, we came down river.'

'And there are horses there?'

'As many as you need, Colonel.'

Cheung took out his map and examined it quickly, the sergeant peering over his shoulder. He traced a finger along the track leading from Chamdo up over the mountain to Ladong Gompa.

'So that's where they're going,' he said softly and turned to the sergeant. 'A Tibetan name.'

'A monastery, from the sound of it, Colonel,' the sergeant said.

Cheung folded the map and turned to the

wounded soldier from Chamdo. 'How far is the village from here?'

'Five miles, Colonel.'

'Then we've no time to waste.' He nodded to the sergeant. 'We'll march there as quickly as possible and get horses.'

'And the wounded, Colonel?'

'Leave them. We'll send someone from the village.'

He pulled up his collar and started to walk along the iron hard road into the falling snow.

13

The Mountain of God

The snow was a living thing through which they stumbled blindly. Death and the valley had slipped far away and they were alone with man's oldest enemy – the elements.

The hillside was rough, strewn with boulders, and the carpet of snow made the going difficult and unsure. At one point, Father Kerrigan's mount plunged to its knees and Hamid grabbed its bridle, pulling it up again by brute strength.

Janet reined in and Drummond moved up beside her. She was covered in snow and her cheeks were flushed as she smiled down at him.

'How are you doing?'

'Fine and so is Kerim.'

The boy was so swathed in blankets that only his single eye showed, but it sparkled suddenly and Drummond knew that he had smiled.

'These horses are used to this kind of country,' Hamid said. 'Let them choose their own way. They know what they're doing.'

'Do you think we'll find the track?' Drummond said.

'I don't see why not. If we keep climbing on a diagonal line to the east, I can't see how we could miss it.'

They started again, Hamid leading followed by Father Kerrigan, Drummond bringing up the rear. The slope steepened as they moved higher and the full blast of the snow, driven by the wind, hit them as they came out on to the bare mountainside.

At one point half-way up a shelving bank, Janet's horse started to slide. Drummond scrambled forward beating it hard across the rump with his clenched fist and it plunged forward.

It was the snow that showed them the

track, the shape of it clear under the white carpet, zig-zagging up the steep slope beneath them and turning into a narrow ravine about a hundred yards to the right.

When they moved into the ravine, they were sheltered from the wind for a while and climbed upwards, the clatter of hooves against the hard ground echoing between the walls. Gradually, the slope steepened, the walls of the ravine fading into the ground and they came out on the bare mountainside again.

As they climbed, the mountain seemed to rise more steeply, and after another hour they went over the rim of an escarpment and looked across a narrow plateau to where the rock face tilted backwards in great, overlapping slabs, most of which were split and fissured into a thousand cracks.

They moved on, heads down against the driving snow, and after another hour Hamid grabbed the bridle of Father Kerrigan's horse and led it into the shelter of some boulders.

'We'll rest for a while,' he said.

Janet handed Kerim down to Drummond and slipped from the saddle. She wiped the snow from her face and smiled wanly. 'It's cold.'

'Too damned cold,' Drummond said.

Father Kerrigan walked forward stiffly, slapping his arms to restore the circulation. 'I'd better have a look at Kerim.'

Drummond crouched down in the shelter of the boulders and Father Kerrigan knelt beside him and gently parted the blankets. 'God bless my soul, but the child's sleeping.'

'Is he all right?' Janet said anxiously. 'He's warm enough, isn't he?'

'Warmer than any of us in that cocoon.' The old man sat down against the rocks. 'Did you bring the contents of my medical bag?'

Janet nodded and slipped her arms through the straps of the military haversack she'd been carrying on her back. She opened it and took out the Thermos flask of tea she had prepared at breakfast.

'What was it you wanted?'

'Never mind, I'll find it for myself.'

The old priest looked grey and haggard and the lines in his face scoured deep into the flesh. He searched through the contents of the haversack and found what he was looking for, a small bottle of red capsules. He slipped a couple into his mouth and Janet passed him tea in the one tin mug that she had brought.

Father Kerrigan took a mouthful down and leaned back with a sigh. Hamid said anxiously, 'Are you all right, Father?'

The old man opened his eyes and grinned. 'Let's just say I'm not as young as I was, but the pills I've taken start acting straight away. I'll make it. The luck of the Irish.'

The mug came round in turn and when it reached Drummond, he swallowed the hot tea gratefully. Hamid produced a couple of cheroots from one of his breast pockets and they lit them and moved away from the others, looking back down the track into the snow.

'The old man doesn't look too good,' Drummond said. 'How long till we reach the monastery?'

'Maybe three hours,' Hamid said. 'It all depends on the state of the track.'

'I've been thinking,' Drummond said. 'What

guarantee have we got that there will be anyone there when we do reach the place? It could have fallen into disuse years ago. There are ruined monasteries all over the mountains, you know that as well as I do.'

'At least we'll find some sort of shelter,' Hamid said. 'And that's something we're going to need just as soon as we can find it. It's no use pretending the old man or Janet and the child, for that matter, can stand much of this sort of thing.'

They moved back to the others and Father Kerrigan got to his feet. Whatever he had taken had certainly had a miraculous effect and he smiled, cheeks slightly flushed.

'I'm ready when you are.'

Hamid helped him into the saddle, Drummond passed the boy up to Janet and they moved on, skirting the base of the great face of rock slabs.

Over the years, the track had been marked by pilgrims placing their stones on conical cairns which marked quarter-mile intervals and these were still clearly visible in the snow.

An hour later, the track turned into a

narrow ravine that slanted up into the rock. It was choked with boulders and loose stones, an indication of years of neglect.

Hamid took the lead, holding Father Kerrigan's horse by the bridle and Drummond did the same for Janet. He was soon tired and his arm ached with the strain of holding in the unwilling horse. He constantly slipped on the snow, sending loose stones rattling through the maze of boulders below.

Once or twice when they paused, he looked up at Janet and was shocked at the weariness in her eyes. Somehow she managed to smile and he smiled back.

Half an hour later they emerged from the ravine on to a ledge perhaps forty feet across that slanted upwards to the left, jutting out from the cliff face.

Hamid turned, still holding on to the bridle of the old man's horse. 'Everyone all right?'

Drummond glanced up at Janet and she nodded. 'Fine. Keep going.'

The ledge lifted steeply, following the curve of the wall and a sea of swirling snow cloaked the valley below. Drummond followed Hamid and Father Kerrigan, holding the horse as close in to the wall as possible.

And then the ledge narrowed until there hardly seemed room for man and animal together. He pushed forward frantically and came out on the edge of a great plateau.

Beyond them, the ultimate peaks of the mountains stabbed into the sky and great sterile valleys ran between, cutting their way through to the other side.

'The main plateau,' Hamid shouted above the wind. 'The monastery can't be very far away. We'll keep on going.'

It was cold at that height, very cold. No more snow fell, but the wind blew harder and harder until it cut through their clothing, whipping their bruised bodies with cold fingers and the child started to cry.

Janet held him close in her arms and Drummond took the reins of the horse, pulling it forward and then they moved over the crest of a small hill and paused.

Below them was a great natural arena into which many valleys spilled, and squarely in the entrance of one of them stood the monastery of Ladong Gompa. Hamid urged Father Kerrigan's mount forward with a savage cry and Drummond went after him.

The monastery walls had been painted red, green and black to signify the nature of the order, but the colours had faded with the years. It was of no great size and had a bleak, deserted look about it. There was no encircling outer wall, a usual feature of larger establishments, and the entrance was at the top of several steps, protected against the weather by a stone porch.

Snow had drifted in an unbroken line across the steps and a chain hung through a hole high in the wall, jingling faintly as it swung in the wind. When Hamid pulled hard on it, a bell rang hollowly somewhere inside and they waited as its brazen sound died.

After a while, they heard a rattle of wooden clogs on stone and a metallic rasping as bolts were withdrawn. The door swung back to reveal a Buddhist monk in faded yellow robes. He showed no particular surprise and came forward at once to give his hand to Father Kerrigan as the old man stumbled up the steps. Drummond held Kerim until Janet had dismounted, then handed him to her and she followed Father Kerrigan.

Another monk, a younger man, came down the steps and Hamid said, 'What about the horses?'

Like the other one, the young man did not speak, but motioned them to follow him and when he tucked his robe into his girdle so that it didn't trail in the snow, Drummond saw that his feet were bare.

There was an enclosed courtyard at the rear. They waited at the gate and after a while it was opened from inside and they moved in. There were the usual stables and a young novice took the horses from them and they followed the other monk into the monastery.

They walked along a narrow, stone-flagged corridor and entered a large, poorly-furnished room at the far end with a fire of logs burning on a large stone hearth.

Janet was sitting by the fire, Kerim nursed in her arms, while Father Kerrigan sat on a bench by a large wooden table, engaged in animated conversation in English with a much older monk in a yellow, conical hat with ear-flaps.

Father Kerrigan got to his feet and the monk rose with him. 'Major Hamid and Mr Drummond.' He made the introductions in English. 'This is the Abbot of Ladong Gompa. I've been giving him a brief account of our misfortunes. Apparently they still get a few pilgrims across during the summer. Lucky for us, eh?'

'I suppose we're pilgrims in a sense,' Drummond said. 'Pilgrims of hope.'

The Abbot smiled. 'I've been explaining to Father Kerrigan that the other members of our order here are under a strict vow of silence. Please accept that they mean no discourtesy.'

His English was slightly stilted and technically excellent, but was delivered in the grave, expressionless tone of a man who did not use his voice often.

'Can we stay here for a while?' Drummond said.

'As long as you wish.'

'Has Father Kerrigan told you that we are being followed by Communist troops?'

The Abbot nodded. 'Sound travels great distances at this height. We could hear your party coming when you were still crossing the main plateau. There will be ample warning. I will have food sent to you and then blankets. I suggest you all try to get some sleep.'

'And that's the most sensible thing I've heard in a long time,' Drummond said.

'I shall pray for your continuing good fortune.'

The Abbot left the room. Hot food was brought to them, steaming in a great copper bowl, and afterwards blankets.

Drummond draped one over his shoulders and Hamid spread the map out on the table. 'Where do we go from here?'

Hamid ran his finger along another valley, following the track over the top and down the other side of the mountain. 'About fifteen miles to the Indian border from here, that's all.'

Drummond looked across to where Father Kerrigan and Janet were already asleep in front of the fire wrapped in their blankets, Kerim between them.

'Do you think they can make it?'

'They'll have to. We don't have any choice.'

He lay down on the floor beside the others, pulling his blanket over his head, and Drummond stayed at the table. It was peaceful, quiet after the storm, the regular breathing of the sleepers rising and falling gently and after a while he rested his head in his arms and slept.

He awakened suddenly, yawned and stretched his arms so that the blanket fell from him. As he bent down to retrieve it he became aware that the Abbot was standing just inside the door watching him.

'How long have I been asleep?'

The Abbot came forward and sat on the bench on the other side of the table. 'About three hours. It is almost night.'

Drummond glanced across at the others sleeping quietly beside the fire. 'They're very tired. They've been through a great deal.'

The Abbot nodded and brooded quietly, face expressionless and calm as the firelight played across it. Drummond felt completely rested and wide awake, but his feet pained him and the toes on his right foot were numb and lifeless.

He fumbled half-heartedly with the laces of his combat boots, but the knots were swollen and tightened by the constant damp of the past two days and he finally gave up trying.

'It would interest me to know what you think of my country,' the Abbot said.

'Frankly, I can't get out fast enough. I've seen enough of places like this, smoke rising from burning cities, refugees on the move.'

'But you came by choice in the first place, did you not?'

'I once read somewhere that life is action and passion,' Drummond said. 'That if a man failed to take part in it, he wasn't really living.'

He absentmindedly banged his right foot against the floor in an endeavour to restore the circulation and the Abbot said, 'A mistake to take that too literally. It was said by a man who, having experienced the horrors of war, devoted himself to the rule of law for the rest of his life.'

The Abbot crossed the floor and opened

a pair of large wooden shutters revealing the night and the mountains. Drummond joined him on a small stone terrace.

It was very cold and he pulled his blanket more closely about him and shivered. During the past few days, his body had been alternately wet and frozen so many times, that he was now at a stage where his resistance was at a very low ebb.

Night was beginning to fall, cold and clear with great scatterings of stars, brilliant and luminous, strung away across the peaks. As he looked, it darkened quickly from east to west and the stars were blotted out before his eyes as though someone moved among them quickly, snuffing them out between finger and thumb.

'It will snow very heavily soon,' the Abbot said.

A small wind lifted the hair on Drummond's head as it skidded round the corner of the building. Gradually, the shadow moved across the night sky until there were no more stars to be seen and the wind howled mournfully as it sped down the valleys towards them.

'It isn't a night I'd like to be out in.'

The Abbot lifted a hand, motioning him to silence. Drummond strained his ears, but heard nothing. He was about to speak when quite suddenly, as the wind lifted, there was a faint jingling sound.

'They are coming,' the Abbot said simply.

'Are you sure?'

The Abbot nodded. 'Crossing the main plateau.'

'Is there anywhere we could hide?'

The Abbot shook his head. 'This is a small place, not like some. As they are looking for you, they will search thoroughly.'

Drummond dropped his blanket, moved to the fireplace and shook the others awake quickly.

Hamid sat up at once. 'What is it? Trouble?'

Drummond nodded. 'We're about to have company. We'll have to get moving again, I'm afraid.'

'I will have your horses made ready,' the Abbot said and he hurried out.

As Father Kerrigan and Janet got to their

feet, Hamid and Drummond moved across to the shutters. Hamid opened one and peered out. He closed it, his face grim. 'It's snowing again. How long are we going to last in the open on a night like this?'

Drummond turned to Father Kerrigan and Janet, standing by the fire. 'If we stay, Cheung will catch us, there's no doubt of that. He'll take this place apart looking for a hiding place.'

'That's all right, Jack,' Father Kerrigan said in a tired voice. 'It isn't your fault.'

The door opened and the Abbot came in with one of the monks, bundles of sheepskins in their arms. 'A sheepskin coat for each of you. Our shepherds find them very useful at this time of the year.'

As they pulled them on, Hamid said urgently, 'Is there anywhere we can go, anywhere at all? We won't last long on a night like this.'

'I think I can help you,' the Abbot said. 'I'll show you as you leave.'

Kerim was still asleep. Janet lifted him gently in her arms and the Abbot led the way along the dark corridor to the courtyard at the rear.

A monk brought the horses forward and helped Father Kerrigan and Janet into the saddle. They all crossed to the gate and the Abbot moved outside with them.

He pointed to the valley beyond. 'This is the best way, the only way. Eight miles and you're through to the other side of the mountain. You'll find a shepherd's hut at the end with wood for a fire, a lantern, everything you need. From there into the valley is easy. Five miles from the mountain and you will come to an Indian border post.'

Powdery snowflakes were already beginning to stick to their sheepskins as the small cavalcade moved away, Hamid leading Janet's horse, Father Kerrigan behind.

'Thanks for everything,' Drummond said.

The wind lifted snow around his legs as he walked away and the Abbot called quietly, 'Do not worry, my friend. You will reach India.'

The snow began to fall steadily till it filled the night and they were alone with it.

* * *

As they advanced towards the end of the narrow valley, the going became heavier and Drummond's feet sank ankle-deep into the snow. He walked with his head bowed against the wind, alone with his thoughts, and when a sharp stab of pain cut into his face, he winced and came to a halt.

To his surprise, he found that he was knee-deep in snow. When he wrenched off a mitten and touched his face, he felt caked snow and ice on his cheeks and his flesh had split in several places. He frowned and pulled on his mitten, and when he looked up saw that he was alone.

The wind was whipping the snow into a frenzy and it spun around his head and sliced at his cheeks, until his face was so numb he could feel no pain.

How long since they had left the monastery? *An hour? Two hours?* There was no knowing, and as a horse whinnied somewhere near at hand he blundered forward.

He peered down at the ground and saw great slurred hoofprints leading away through the snow and stumbled forward, half-bent so that he could follow them.

Time had stopped and his frozen mind had difficulty in thinking what to do next. The wind was howling like a lost thing and he was completely covered with frozen snow until he no longer resembled an ordinary man. He fell several times, and each time lay in the snow for a little longer before getting up.

A terrible iron band settled around his chest and he seemed to be struggling for breath. Again he heard the whinny of a horse and then it appeared from the whirling darkness, rearing up above his head, Father Kerrigan falling over the hindquarters and knocking him to the ground.

As Drummond sat up, the wind carried the sound of the horse's desperate cry and there was a coldness sweeping into his face, a sense of space, of limitless distance. He crawled forward, feeling the ground in front of him and then his hand touched nothing but air.

He crawled backwards, turned and went back to the old man. Father Kerrigan was on his hands and knees like an animal, his body coated with snow, and Drummond heaved him to his feet and they staggered forward.

It was no good. He was on his knees, the old man beside him in the snow, his arms moving feebly. Drummond took a deep breath, something deep inside, some essential courage that refused to be beaten giving him the strength to haul the old man to his feet.

They stood there, swaying together and then the other horse loomed out of the night, Hamid in the saddle.

What happened after that was something Drummond could never really remember afterwards. He was aware of Hamid pulling the old man up across the saddle with a supreme effort, of shouted directions that were snatched away by the wind and then the horse plunged forward, taking him with it, his right hand hooked firmly around the saddle girth.

It was Janet at the door of the hut with the lantern that saved them and the light drew them out of the storm. Hamid slid to the ground, pulling Father Kerrigan after him and staggered towards the door while Drummond hung on to the horse.

It was no use. As a sudden gust of wind slashed in from the valley driving razor-sharp particles of ice before it, the terrified animal reared up, knocking Drummond to the ground and galloped madly into the night.

He was on his hands and knees again, crawling towards the doorway and the wind seemed to have got inside his brain, dragging him down into the whirling darkness.

14

The Last Round

He awakened slowly and lay for several moments staring up through the gloom, trying to decide where he was. Realisation came suddenly and completely and he tried to sit up.

The hut was low roofed and built of blocks of rough stone. He was lying on a pile of mouldy hay with Hamid beside him. In the middle of the floor a fire burned brightly.

All his outer clothing had been removed and he was only wearing his underwear. He had been covered with sheepskin coats and he pulled them aside and examined his swollen, chapped hands. Gingerly, he touched his face and winced as fingertips probed great splits in his flesh.

His right foot felt heavy and numb and when he sat up, he saw that it had been bandaged. He reached to touch it and Hamid opened his eyes and pushed himself up on one elbow.

'How do you feel?'

'Bloody awful. What's wrong with my foot?'

'A touch of frostbite, nothing serious. All your toes are still there, if that's what you're thinking.'

'I can't feel a damned thing.'

'Janet gave you an injection. Something from the old man's medical kit.'

Drummond looked across to the other side of the fire to where Janet, Father Kerrigan and the young Khan slept peacefully. 'How is he?'

'He had a heart attack when I got him inside last night. Luckily he'd brought the right sort of drugs along and Janet was able to give him an injection.'

'He's in pretty bad shape then?'

'Couldn't walk another step and, in case you don't remember, we lost both horses last night.'

He took a cheroot from his pocket, broke it in two and handed Drummond half. 'The last one so make the most of it.' He walked to the door, opened it slightly and peered out. 'Dawn's coming and the snow seems to be lifting.' He returned to the fire and pulled on his boots. 'I'll take a walk and find out exactly where we are.'

The door closed behind him softly and a small, trapped wind raced round the walls seeking an outlet and then died. There was a sudden movement in the shadows on the other side of the fire and Janet sat up.

'Jack, are you all right?'

'Fine,' he said softly. 'Ali's gone to have a look round.'

He started to dress, fumbling over the buttons with his swollen fingers and she threw some more wood on the fire. 'How's your foot?'

'I'm just beginning to feel it again.'

'I'd better give you another injection.'

He was hardly aware of the needle going in. 'How long will that last?'

'Four or five hours.'

He found the laces of his combat boots quite impossible and she tied them for him after fitting the right boot over the bandaged foot gently.

'How's that?'

'Fine.' He took her hand. 'You look just about ready to fall down. How's Father Kerrigan?'

'Not too good, I'm afraid. He needs hospital treatment.'

'And Kerim?'

She chuckled. 'In better shape than the rest of us put together, I think.'

There was a sudden draught as the door opened, then closed and Hamid dropped by the fire, cursing softly and holding his hands to the flames.

'What's it like?' Drummond said.

'Cold enough to freeze you to the ground, but it's stopped snowing.'

'What about getting out of here?'

'We're in the hollow of a small plateau overlooking the lower slopes of the mountain. According to the Abbot, it's five miles down to the big valley and the Indian border.'

'How rough is the going?'

Hamid shrugged. 'Impossible to tell, it's not quite dawn yet, but it shouldn't take us more than a couple of hours even if conditions are bad. It's all downhill.'

Drummond got to his feet and swayed slightly, suddenly light-headed. 'Are you all right?' Janet said anxiously.

He nodded and walked carefully to the door. Outside it was still dark, but towards the east, a pale, grey light was lifting over the peaks. He followed the line of Hamid's footsteps, climbing up out of the hollow and stood on the rim, looking down into the darkness of the valley.

After a while, he turned and went back to the hut. Hamid glanced up at him as he dropped beside the fire.

'Well, what do you think?'

'The old man will never make it.'

'We could carry him.'

Drummond shook his head. 'We'd have enough trouble getting ourselves down there on foot. Even Janet would find it a struggle.'

'Then what do we do?' Hamid said. 'We can't leave him.'

There was a tired chuckle from the other side of the fire and Father Kerrigan said in a faint voice, 'You haven't any choice, have you?'

'I'm damned if I will,' Drummond said. 'If we assume that Cheung and his men stayed overnight at the monastery because of the storm, then we've got to expect that he'll start out again at first light, especially as it's stopped snowing. He's come this far, he won't stop till he reaches the border and has to accept the inevitable.'

'So what do we do?' Hamid said. 'Stay here and try to beat him off?' He picked up the Garrand. 'With one rifle.'

'What's your suggestion?'

'If we got to the border post fast enough, we could get help.'

'And come straight back?'

'That's right. For all we know, they may have air support down there, helicopters even. They're bound to be reinforcing the entire area in view of what's happened.'

Drummond stood there, indecision on his face and Janet said quietly, 'He's right, Jack, it's the only plan that makes any sense. I'll stay here with Father Kerrigan.'

'Now just wait a minute . . .' Drummond began.

She shook her head, her face grave. 'I'm staying, Jack, he needs me, but you must take Kerim with you.'

'But why, for God's sake?' Drummond demanded. 'We'll be coming back for all of you.'

'You may not be in time.'

She stood before him, arms hanging straight at her sides, calm and determined, her eyes very tired, and then she smiled and there was all the love in the world there for him.

'Hurry back, Jack! Hurry back!'

He reached blindly for her and Hamid took him firmly by the arm. 'We're wasting time, Jack.'

Drummond turned and stumbled to the door and Hamid offered her the rifle. 'I'll leave you this.'

She shook her head. 'I couldn't use it, Ali,' she said simply.

Hamid stood there for a moment, a frown

on his face and then he slung the rifle over his back and went round the fire to where Kerim slept beside the old priest, swathed in his blankets.

He picked the boy up gently, cradling him in his arms and Father Kerrigan smiled. 'I'd take it as a personal favour if you'd run all the way, Major.'

Hamid turned and went out, the lump that rose in his throat threatening to choke him. Drummond was waiting outside and the Pathan walked past him without speaking, the boy held close to his chest.

Drummond stumbled after him. On the rim of the hollow he paused to look back at the hut. Janet was standing in the entrance. She gazed towards him for a long moment and then went back inside. The door closed with a strange finality and Drummond turned and went down the slope after Hamid.

Progress was slow at first for on the upper slopes, sheltered by a shoulder of the mountain, the snow had not been swept away and had fallen in a deep blanket that made walking difficult.

Drummond soon realised how weak he was. They had not covered a mile before he was gritting his teeth and placing one foot in front of the other with a dogged persistence. Hamid seemed tireless and ploughed ahead through the snow without faltering, but his face, when they rested in the lee of a large boulder, told another story.

Kerim's single eye over the edge of the blanket was round with wonder and Hamid laughed. 'I wonder how much of this he'll remember in the years to come?'

'God knows,' Drummond said hoarsely. 'Here, give him to me. I'll take him for a while.'

Hamid didn't even try to argue, a bad sign, and they started to walk again. The boy seemed heavy, which was a strange thing, and Drummond held him close and leaned well back as he went down the slope.

Another mile and his legs were trembling and when he tried to take another pace forward, he overbalanced and rolled over and over down the mountainside.

He held on tight to the boy and the world spun and red sparks flashed before his eyes. Faintly, through a great roaring, he heard Hamid calling to him and he came to rest in a great drift of snow.

The boy was crying and Hamid picked him up and brushed snow from his face as Drummond got painfully to his feet. Hamid's eyes seemed to have receded into their sockets, and lines of fatigue were etched deeply into his face. They didn't speak – there was nothing to say. He started to march, the boy against his chest and Drummond followed.

Time no longer had any significance for Drummond. He placed one foot doggedly in front of the other, and after a while they left the slopes and struggled over a flat plain of deep snow. Half-way across, they had to rest, completely exhausted.

Darkness had fled across the mountains and day had dawned, grey and sullen, more snow threatening in the heavy clouds as they finally struggled out of the deep snow and entered a thinly wooded stretch that sloped down to the valley bottom.

Drummond sucked a piece of ice, delighting in the coolness of it as it melted in his mouth and trickled down his throat and hobbled along in a strange, trance-like mood.

It was with a sense of shock that he found himself lying in the snow, the taste of it cold in his mouth and then a foot dug into his side and he heard Hamid's dead, washed-out voice.

'Get up, Jack. I haven't the strength to lift you.'

He turned away and Drummond with a supreme effort got to his feet and went after him. He bowed his head and placed one foot in front of the other. He repeated that simple action until he had lost count of time and suddenly heard a shout in front.

Hamid had stopped on top of a slight rise twenty or thirty yards away and called to him in a strange, cracked voice. Drummond broke into a stumbling run and reached the top of the rise in time to see Hamid staggering down towards the camp in the hollow below. There were field guns deeply entrenched, supply trucks parked at the rear and a sprinkling of snow-covered huts.

Men were flooding forward, men in familiar uniforms and khaki turbans, some riding supply mules. They reached Hamid and Drummond saw him hand the boy carefully to a great, bearded Sikh. He turned, looked back at Drummond, took a single hesitant step and fell on his face in the snow.

Drummond slid to the ground and sat there, tears rolling down his cracked cheeks as the soldiers moved towards him.

It was warm in the hut and he sat before the stove, a blanket round his shoulders and sipped hot tea slowly, holding the mug in both hands. After a while, the door to the other room opened and a young Bengali medical corps sergeant came in.

'How is he?' Drummond asked.

'Fine,' the sergeant said. 'He's fallen asleep now, quite exhausted.'

'And the boy?'

'Having a meal in the officers' mess, such as it is.' The sergeant laughed. 'There's nothing wrong with that one. He seems to have enjoyed himself, if anything, during the past few days. More brandy?'

Drummond nodded and held out his mug. 'How much longer will your commanding officer be?'

'He shouldn't be long now. The main command post is only three miles away, but since the snow, of course, we're having to use mules.'

The door swung open, a cold wind whistling round the room and young Lieutenant Singh entered. 'Major Naru's coming now, Mr Drummond.'

'Thank God for that.'

Drummond got to his feet and hobbled to the window in time to see the major and an escort of two privates ride up on mules. They dismounted and the major came up the steps to the hut, brushing snow from his parka with both hands.

Lieutenant Singh opened the door for him and he came in and moved straight to the fire, a tall, handsome man with a clipped moustache.

'Mr Drummond?' He pulled off his gloves and held out a hand. 'A pleasure to see you here, sir.'

'Believe me, it's a pleasure to be here, Major,' Drummond said. 'Did Lieutenant Singh give you the whole story?'

The major nodded. 'We spoke over the field telephone. Where is Major Hamid?'

'Asleep in the other room. He's done the work of ten men during the past few days.'

'And the young Khan?'

'We're looking after him in the officers' mess, sir,' Lieutenant Singh put in.

'What about my friends, Major?' Drummond said. 'When can we make a start? I wanted to return with men and mules straight away, but the lieutenant said he couldn't move without the good word from you.'

Major Naru sighed. 'I'm afraid it's rather more complicated than that. The Chinese invasion of Balpur is something my government must handle with the greatest care. An emergency session has already started at the United Nations. Under these circumstances, all units on the border have been ordered to avoid any confrontation with Chinese units at whatever the cost. It would be impossible for me to even consider sending a patrol into Balpur territory.'

'But that hut's no more than five miles from here,' Drummond said. 'With mules, we could be there in less than an hour and time is vital. As I explained to Lieutenant Singh, Colonel Cheung could beat us to the punch.'

'All the more reason to avoid a situation which could lead to possible military action.'

'We'll see what Major Hamid has to say about this,' Drummond said angrily and he moved to the door of the inner room.

'Major Hamid is an officer of the Indian Army. He will do what I have to do – obey orders.' Major Naru's voice cracked suddenly. 'Do you think I'm enjoying this, Mr Drummond? If I had my way, I'd move over that border now with every man I've got.' He pulled on his gloves. 'I'm going to get in touch with Headquarters by radio immediately. If they give me the word, I'll lead my men in myself, I promise you.'

'How long will that take?'

'To get a reply?' Major Naru shrugged. 'An hour, perhaps two. It is something they will have to consider carefully.' He moved to the door and Singh opened it for him. 'I am sorry, Mr Drummond.'

The door closed behind them and Drummond went to the window. Major Naru walked across to the command post, Singh at his shoulder. The three mules he and his escort had used were tethered outside. Drummond looked at them for a moment, then made his decision.

The medical sergeant was standing by the stove, his face troubled, and Drummond moved past him and opened the door to the inner room. Hamid lay on his back on one of the bunks, breathing gently, the harsh lines smoothed from his handsome face.

When they had carried him in, someone had brought his rifle and it stood in the corner by the window. Drummond slung it over his back and looked down at Hamid for a moment.

'Good luck, Ali,' he said softly and returned to the other room.

He uncorked the brandy bottle, poured some into his mug and swallowed it quickly and the medical sergeant watched, a frown on his face.

'Why the rifle, Mr Drummond?'

'I'm going for a little ride,' Drummond said. 'It might come in useful.'

He went to the door and opened it, the sergeant hurrying at his shoulder. 'But this is madness.'

Drummond ignored him, went down the steps pulling on his mittens and crossed to the mules. As he unhitched them, the medical sergeant ran past him, mounted the steps to the command post and went inside.

Drummond took his time, lopped the reins of two of the mules to the pommel of the saddle of the third, mounted, and rode away.

He passed between the field guns, men standing up to stare at him, and then Major Naru and Lieutenant Singh emerged from the command post, the medical sergeant at their backs, and hurried after him.

As Drummond passed the last gun emplacement, they caught up with him and Major Naru reached for the bridle of the mule he was riding.

'I can't let you do this, Mr Drummond.'

'Then you'd better start shooting,' Drummond said calmly. 'It's the only way you're going to stop me.'

He jerked the bridle from the major's grasp, dug his heels into the mule's flanks and moved forward. When he reached the crest of the small hill and looked back into the hollow, Major Naru was still standing there in front of the gun, but Lieutenant Singh was running back towards the command post.

The clouds had dropped down towards the jagged peaks, heavy with snow, and as the mules moved out of the valley and started up the mountain, the first few flakes started to fall.

Drummond no longer felt tired, but there was a strange singing inside his head, perhaps the brandy talking, and he was alone in a great white silence, following the double track in the snow that he and Hamid had made on their way down.

He pushed the mules as much as he could, moving up into the white stillness towards the peaks as the snow continued to fall. It was just under an hour after leaving the camp that he came out of a ravine on to the final slope and moved up towards the plateau.

From the rim of the plateau, sheltered by a group of jagged rocks, Sergeant Ng watched his progress from the moment he emerged from the ravine. As Drummond drew closer, he turned and hurried down to Colonel Cheung who stood beside the horses outside the hut in the hollow below.

Cheung looked tired and the skin of his face stretched tightly over his cheekbones, was raw with frostbite. 'One man coming with three mules,' Sergeant Ng said.

'Take the horses inside,' Cheung told him and he moved up out of the hollow to the rim of the plateau.

He watched Drummond for a full minute

and there was no excitement in his heart. He had failed, utterly and completely, and in Pekin he would have to face the consequences of that failure, but at least he would have something of value to take back with him.

He ran down into the hollow and went inside the hut. The horses had crowded to the far end and were quietly feeding on the hay. Father Kerrigan was sitting up on the other side of the fire. Janet standing beside him and Ng waited by the door.

'It's Drummond,' Cheung said. 'I'll stay down here. You wait for him in the rocks on the edge of the hollow. Let him ride past you before you make your move.'

'Do you want him alive?' Ng asked calmly.

'At all costs.'

Ng went out, closing the door behind him and Cheung drew his revolver. He smiled gently across the fire at Janet and Father Kerrigan.

'It would be unwise for either of you to attempt to make the slightest noise, do I make myself clear?'

* * *

Drummond came over the edge of the plateau and reined in. It was a peaceful scene, the hut standing below in the hollow, smoke rising into the gently falling snow. He had unslung the Garrand as a precaution while still in the ravine and now it rested across the saddle in front of him.

He dug his heels into his mule's flanks and started into the hollow. He was perhaps half-way down the slope, when there was a commotion inside the hut, the door was flung open and Janet ran outside.

'Behind you, Jack!' she called. 'Behind you!'

Drummond released the two lead mules and jerked savagely on the bridle of his own mount, pulling it round as Sergent Ng emerged from the rocks at the top of the hollow, sub-machine gun in his hands.

He fired a warning burst into the air and Drummond's mule reared, throwing him over its hindquarters as he reached for the Garrand.

He came to his knees in deep snow, the Garrand still in his hands, the three mules milling around him. Sergeant Ng crouched, trying to get a clear view, and Drummond fired twice in rapid succession, the bullets somersaulting the Chinese back over the rocks.

As the mules broke away, trotting down to the hut, he turned and saw Janet on one knee, Cheung holding her by the hair, the barrel of his revolver rammed against her neck.

Drummond walked forward, the rifle at his hip, and stopped a yard or two away. 'Let her and the old man go, Cheung, take me. I could be of real value, more than you could ever realise.'

'No bargains, Jack, quickly now.'

Cheung's voice was as soft as ever, but quite implacable, and as he thumbed back the hammer of his revolver, Drummond threw the Garrand far away into the snow.

'That's better.'

As Cheung released his hold, Janet came to her feet and ran into Drummond's arms. He held her close for a moment. 'Is Father Kerrigan all right?'

She nodded. 'What about Ali and Kerim?'

'We reached the border safely, but the Indian Army has strict orders about crossing over. I had to come back alone.'

'How very fortunate for me,' Cheung said and he pulled Janet back against him. 'You've caused me a great deal of trouble, Jack. I tried to follow you from Ladong Gompa last night and got caught in the blizzard. We had to turn back. Only Sergeant Ng and I made it. I knew I'd be too late and yet I still came on this morning. That's the kind of man I am.'

'Father Kerrigan and the girl can't be of any use to you. Let them go. I won't give you any trouble.'

'They want you in Pekin, Jack,' Cheung said. 'They know all about the work you've been doing for Ferguson. When you stand before a military tribunal, these two will stand beside you, I'll see to that personally.'

Drummond shook his head. 'You'll be facing your own tribunal, Colonel. You lost the young Khan, remember?'

Something glowed in Cheung's eyes, he

pushed the girl away from him and the revolver came up. Drummond tensed himself to spring, knowing already that he was too late.

Cheung took a deep breath and shook his head. 'Oh, no, Jack, nothing as easy as that, I promise you.'

Somewhere horses plunged and snorted and a hard, familiar voice called, 'Over here, Cheung!'

Hamid was already sliding from the back of a mule up on the rim of the hollow, an automatic rifle in his hands. Cheung turned, crouching, and Hamid fired three times so quickly that they sounded like one, the first shot catching Cheung in the shoulder, spinning him round, the second and third driving him hard against the wall.

Janet turned away quickly, stumbling into Drummond's arms as Cheung struggled for life, clawing for the revolver he had dropped, and then blood erupted from his mouth in a bright flow and he coughed once and lay still.

Lieutenant Singh came over the skyline on a mule, reaching for the bridle of Hamid's mount, following him down into the hollow. Hamid turned Cheung over with his toe and looked down at him.

'The face of the damned.'

'What happened?' Drummond said. 'I thought the Indian Army was supposed to stay on its own side of the border?'

'It still is,' Hamid said. 'Young Singh here, woke me just after you left. He thought I might have other ideas, which I did. Being a young man of spirit, he decided to come with me.'

'And Major Naru?'

'Most unhappy.'

'Do I detect the possibility of a court martial in the near future?'

'A matter of supreme indifference to either of us, but, in any event, unlikely. The newspapers wouldn't like it. Is Father Kerrigan all right?'

'As right as he ever will be,' the old priest said, appearing in the doorway. 'Nothing that a bottle of Jamiesons' and a decent meal wouldn't cure.'

'Then I suggest we take you to where we

can obtain both items as quickly as possible. Poor Naru will be most uncomfortable until we cross the border.'

They brought out the horses and helped Father Kerrigan and Janet into the saddle. The old man looked down at Cheung, crossed himself and muttered a prayer as he moved off between Singh and Hamid, and Janet followed.

Drummond was the last to leave, and after he'd mounted, he sat on his mule for a moment or two looking down at Cheung, feeling strangely sad.

But nothing mattered now except that life began again, and as he rode up towards Janet, waiting for him on the edge of the plateau, he was smiling.